COLLATERAL WARRANTIES
A Practical Guide for the Construction Industry

WINWARD FEARON & CO.
David L. Cornes BSc (Eng), AKC, MICE, CEng
Solicitor of the Supreme Court

Richard Winward LLB
Solicitor of the Supreme Court

OXFORD
BLACKWELL SCIENTIFIC PUBLICATIONS
LONDON EDINBURGH BOSTON
MELBOURNE PARIS BERLIN VIENNA

BSP Professional Books
A division of Blackwell Scientific
    Publications Ltd
Editorial offices:
Osney Mead, Oxford OX2 0EL
25 John Street, London WC1N 2BL
23 Ainslie Place, Edinburgh EH3 6AJ
3 Cambridge Center, Cambridge
    MA 021412, USA

First published 1990

Set by DP Photosetting, Aylesbury, Bucks
Printed and bound in Great Britain by
Hartnolls, Bodmin, Cornwall.

DISTRIBUTORS

Marston Book Services Ltd
PO Box 87
Oxford OX2 0DT
(*Orders:* Tel: 0865 791155
        Fax: 0865 791927
        Telex: 837515)

USA
    Blackwell Scientific Publications, Inc.
    3 Cambridge Center
    Cambridge, MA 02142
    (*Orders:* Tel: (800) 759-6102)

Canada
    Oxford University Press
    70 Wynford Drive
    Don Mills
    Ontario M3C 1J9
    (*Orders:* Tel: (416) 441-2941)

Australia
    Blackwell Scientific Publications
    (Australia) Pty Ltd
    54 University Street
    Carlton, Victoria 3053
    (*Orders:* Tel: (03) 347-0300)

British Library
Cataloguing in Publication Data

Collateral warranties: a practical guide for the
    construction industry.
    1. England. Buildings. Defects. liability of
    construction industries. Law
    I. Winward Fearon & Co.
    II. Winward, Richard   III.
    Cornes, David L.
    344.20636

ISBN 0–632–02919–6

Contents

Preface

Since the mid 1980s, the use of collateral warranties has been growing in the United Kingdom. However, the growth rapidly accelerated after the House of Lords' decision in *D & F Estates Ltd and Others* v. *The Church Commissioners for England and Others* in 1988 and will accelerate even more rapidly since the further House of Lords' decision in *Murphy* v. *Brentwood District Council* in July 1990. These cases have had the effect of substantially removing the prospects of tenants and purchasers bringing successful actions in tort in respect of the consequences of defective design and construction in the building industry. In practice, this means, for example, that a leaseholder of a defectively designed commercial office building will be unable to bring an action in tort against the architect who had been employed by the developer. Collateral warranties seek to fill that gap by creating a contractual relationship between the parties.

One of the side effects of the enormous growth in the use of collateral warranties has been the attempts by each of the parties involved to try to have their own interests dominate in the negotiation process. Like all contracts, collateral warranties seek to allocate risk and developers, banks, pension funds, contractors, architects, engineers, quantity surveyors, tenants and purchasers each try to take as little risk as possible and seek to protect their own interests. Inevitably, this has produced a plethora of non-standard forms of warranty, some as short as one page and some occasionally reaching 40 pages or more. Some have been drafted by commercial conveyancing lawyers who appear to have little understanding of the complex interaction of collateral warranties with various issues such as the contract to which they are collateral (JCT 80, Architect's Appointment, the ACE Conditions and so on), limitation of action, insurance and the whole basis of professional liability. Many collateral warranties have of course been drafted by lawyers who do understand the issues and of those, some are a sensible commercial compromise between competing vested interests and some are not.

These vested interests have made the production of standard forms of warranty very burdensome for those who have tried. There is, at the time of writing this book, only one standard form in England and Wales: 'the Form of Agreement for Collateral Warranty for use where a warranty is to be given to a company providing finance for a proposed development' produced by the British Property Federation, the Association of Consulting Engineers, the Royal Institute of British Architects and the Royal Institution of Chartered Surveyors. This form took three years to agree. Even before its launch in May 1990, there were voices of criticism being raised in public including that of a former Chairman of the RIBA's liability committee. The difficulties that the BPF, ACE, RIBA and RICS face in trying to agree further collateral warranties for tenants and purchasers cannot be underestimated. We hope that those difficulties may be reduced by the contents of this book and the constructive criticism of the existing warranty which is in Chapter 9. In a similar way, the difficulties faced by the Joint Contracts Tribunal in trying to agree, as they are, warranties for contractors and sub-contractors are enormous.

Our purpose in writing this book has been to try to explain the law behind collateral warranties and the issues that arise in drafting and considering them so as to demystify the subject and to clear away some widely held myths. We hope that we have succeeded. Whilst we have sought to explain the law and the issues as we see them, a book is no substitute for good legal advice in the area of collateral warranties.

We have tried to state the law in England and Wales as at 31 August 1990. Although the basis of the law is that of England and Wales, we hope that readers in Scotland and Northern Ireland will find the text helpful.

We are grateful to Ann Hockin with help from Wendy Burns, Valerie Preston and Kay Oliver who have typed the text (without any complaint) and to all those who encouraged us to embark on and continue with this project, including Julia Burden, our commissioning editor from Blackwell Scientific Publications Ltd. We are also grateful to our partners for their help, advice, encouragement, patience and constructive input.

We are pleased to acknowledge the kind permission of the British Property Federation, the Association of Consulting Engineers, the Royal Institute of British Architects and the Royal Institution of Chartered Surveyors, who jointly own the copyright, to reproduce the Form of Agreement for Collateral Warranty (CoWa/F). Butterworth & Co. (Publishers) Limited kindly gave permission to reproduce some extracts from Volume 22 of *The Encyclopedia of Forms and Precedents, Landlord and Tenant: Business Tenancies*, which sets out some examples of typical repairing covenants in leases (these appear in Chapters 6 and 10). The

Royal Incorporation of Architects in Scotland gave their kind permission to reproduce the RIAS approved duty of care agreement in which they hold the copyright — that form is for use under the law of Scotland.

Winward Fearon & Co.
35 Bow Street
London WC2E 7AU

David L. Cornes
Richard Winward
August 1990

Chapter 1

Principles of Law

DEFINITION OF COLLATERAL WARRANTY

Construction projects, by their very nature and also by their method of **1.1**
procurement, create complex legal relationships between the many
parties involved in the design and construction process. Further the
finished product has, or should have, a life expectancy calculated not in
years but in tens of years giving rise to a class of future owners who had
no involvement in nor control over the original contract works but who
may have substantial liabilities if things go wrong. It is because of these
characteristics that tort was a useful remedy permitting a much wider
forensic examination of liability and a greater potential for remedy and
indeed disputes. Tort also permitted a more flexible apportionment of
blame amongst the parties involved in the actual construction process i.e.
employer, architect, engineer, quantity surveyor, main contractor, sub-
contractor or supplier, sub-sub-contractor and the building control
authority. The use of tort as a remedy in construction cases has been
substantially eroded for the reasons set out in Chapter 2. In its place there
is now widespread use of a contractual remedy in the form of collateral
warranties.

In contract the role of the court is restrictive unlike tort where its role **1.2**
is creative. In contract, subject to statutory exceptions, the basic role of
the court is to police a set of rules to ensure that all the component parts
of a binding contract have been satisfied; the nature and substance of the
bargain between the parties is a matter of private agreement provided of
course that bargain is not illegal nor contrary to statute or public policy.
In tort the duty and standard of care is created by the courts and is not a
matter of private bargain.

A basic rule of contract is the doctrine of privity of contract. This **1.3**
doctrine states, as a general rule, that only a party to a contract can take
the benefits of that contract or is subject to its burdens or obligations. For

example if A promises to B to pay a sum of money to C, as a general rule, C cannot enforce that obligation against A.

1.4   A collateral warranty is one of the ways of overcoming the restriction on remedies created by the doctrine of privity. In law the term 'collateral warranty' has several meanings. It may mean a warranty or representation which is collateral to the main transaction: *De Lassalle* v. *Guildford.* L had agreed the terms of a lease with G. However L refused to complete the transaction unless and until he had been given an assurance by G that the drains of the property were in good order. G gave an appropriate verbal assurance and L completed the lease document. After going into possession of the property L found that the drains were defective. He was not able to bring an action under the terms of the lease. However he claimed damages against G on the basis of a collateral warranty created by G's verbal assurances. L's action succeeded.

1.5   A collateral warranty which is collateral to the main transaction will give an additional cause of action to one of the parties to that transaction but does not introduce a third party. However the term collateral warranty has a further meaning which is more apposite to the subject matter of this book, namely a binding contract entered into between B and C which is collateral to a contract already in existence between A and B whereby B promises to C that he will perform his obligations to A. If B is in breach of those obligations to A then C will have a right of action against B. This collateral warranty or collateral contract may be imputed, that is to say arising impliedly as a matter of fact from the circumstances of a particular case or may be created by the parties entering into a specific, usually written, contract. A further species of collateral warranty is a unilateral undertaking entered into by one party in favour of another, the document being a deed.

1.6   In *Shanklin Pier Ltd* v. *Detel Products* S, who were owners of a pier, entered into a contract with G. M. Carter (Erectors) Ltd for the repair and repainting of the pier, the repainting to be carried out with two coats of bitumastic or bituminous paint. Under the terms of their contract with Carter, S reserved the right to vary the specification. D were paint manufacturers who produced a product known as DMU which they represented to S as being suitable for the repainting of the pier in that its surface was impervious to dampness and could prevent corrosion and creeping of rust with a life of 7–10 years. In consideration of this representation S specified to Carter that Carter should use D's paint. The paint proved to be a failure and S sued, not Carter as main contractor, but D as supplier of the paint. D argued that they had not given any such representation or warranty and even if they had, it did not give rise to a cause of action because they were not parties to the contract for repainting the pier. The court rejected this argument and held that on the facts there was a warranty and that the consideration for the warranty was that S should cause Carter to enter into a contract with D for the supply of their paint for repainting the pier. The representations given by D to S were contractually binding in the form of a collateral contract which arose by implication.

In the *Shanklin* case D was aware that their product was to have a   **1.7** specific use by a third party. This was not the situation in *Wells (Merstham) Ltd* v. *Buckland Sand and Silica Co. Limited.* B were sand merchants and they warranted to W, who were chrysanthemum growers, that their BW sand conformed to a certain analysis that would be suitable for the propagation of chrysanthemum cuttings. W ordered a load of BW sand direct from B and a further two loads via a third party who were builders merchants and not horticultural suppliers — that is to say they supplied sand for various purposes. The third party purchased two loads of sand from B for onward sale and delivery to W and gave no indication to B that the sand was required for W or for horticultural purposes. The sand did not conform with the analysis and as a result W suffered a loss in the propagation of their young chrysanthemums. The court held that B was liable on the basis of a collateral contract and that it was irrelevant that the purchase of the sand had been made through a third party; as between a potential seller and a potential buyer only two ingredients were required to bring about a collateral contract namely:

(1) a promise or assertion by the seller as to the nature, quality or quantity of the goods which the buyer might regard as being made with a contractual intention; and
(2) the acquisition by the buyer of the goods in reliance on that promise or assertion.

In contrast to the decisions in *Shanklin* and *Wells* the court in *Drury* v.   **1.8** *Victor Buckland Limited* rejected the plaintiff's claims against the defendant. In this case D was approached by an agent of V in order to persuade D to purchase an ice cream maker. D had doubts as to whether she could afford the equipment. However she was persuaded to pay a 10 guinea deposit to V and to discharge the balance of the purchase price by way of instalments to a finance company. The full terms of the transaction were set out on an invoice sent to D by V. However, the deal was put through by V selling the machine to a finance company who in turn entered into a hire purchase agreement with D. The machine proved

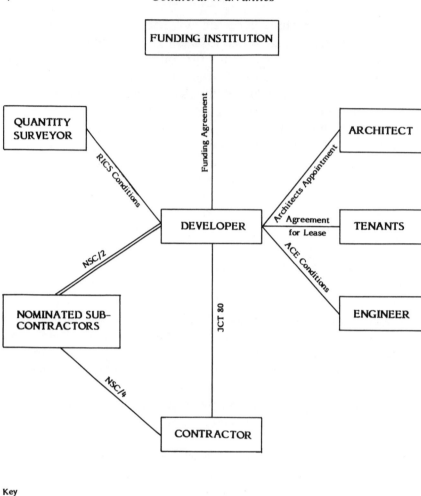

Key

===== Collateral Warranties

Fig. 1: Contractual arrangements arising from an unmodified JCT 80 contract

unsatisfactory and D brought proceedings against V on the basis that V was in breach of the implied warranties or conditions as to quality imposed by the Sale of Goods Act 1893. The court held that there was no contractual relationship between D and V. From the case report it does not appear that D tried to argue that there was a collateral contract. There was evidence before the court that 'in the course of one or other of the interviews in answer to her question "suppose it goes wrong?" he [the defendant's agent] said that if it did she was to ring them up and that she

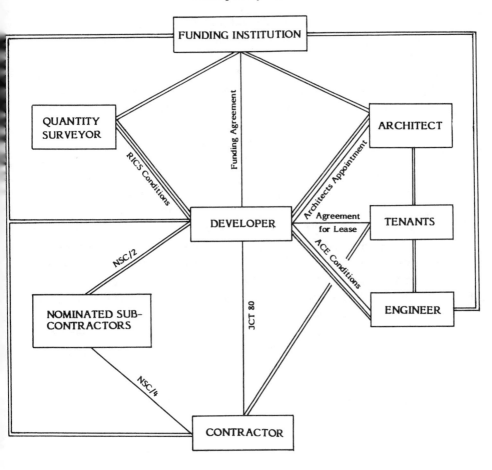

Key

══════ Collateral Warranties

Fig. 2: Typical arrangement after parties have entered into collateral warranties

had no need to worry about it as they would service it for 18 months free'. At the very least it would appear arguable to the authors that those words could have constituted a collateral warranty giving rise to an obligation as between D and V.

A helpful description of the nature of a collateral contract was given by **1.9** Lord Moulton in the case of *Heilbut Symons & Co.* v. *Buckleton* when he stated:

'It is evident, both on principle and on authority, that there may be a contract the consideration for which is the making of some other contract. If you will make such and such a contract I will give you £100 is in every sense of the word a complete legal contract. It is collateral to the main contract but each has an independent existence and they do not differ in respect of their possessing to the full the character and status of a contract.'

1.10 Lord Moulton placed emphasis on the word 'contract'. In the cases of *Shanklin* and *Wells* the courts found that collateral contracts arose by implication from the facts of each case. In the construction industry collateral contracts will invariably arise by the acts of the parties entering into a specific agreement. Regardless of whether the contract arises by implication or by specific agreement a basic understanding of the rules as to the formation and construction of a contract is an essential prerequisite to understanding the problems arising from collateral warranties.

1.11 The provision of a contractual network of collateral warranties for a typical building project creates a network of considerable complexity. Figure 1 sets out in diagrammatical form the contractual arrangements arising from an unmodifed JCT 80 Standard Form of Building Contract. By comparison Figure 2 sets out a typical arrangement after the parties have entered into the collateral warranties which are now commonly sought in respect of a building development.

A COMPARISON OF CONTRACT AND TORT

1.12 The classic ninteenth century definition of contract is 'a promise or set of promises which the law will enforce': Pollock, *Principles of Contract* (13th edition). That is to say, there is reciprocity of undertaking passing between the promisor and the promisee. Tort is generic in nature and therefore more difficult to define. Basically tort is a collection of civil law remedies entitling a person to recover damages for loss and injury which have been caused by the actions, omissions or statements of another person in such circumstances that the latter was in breach of a duty or obligation imposed at law.

1.13 In contract the rights and obligations are created by the acts of agreement between the parties to the contractual arrangement. In tort the rights and obligations are created by the courts applying the common law which has on the basis of previous authority fallen into three distinct categories, namely negligence, nuisance and trespass. Historically both actions in contract and in tort derived from the same source, namely trespass, compared with actions for breach of a deed which were based upon an action on the covenant. Actions for breach of contract were on the case for *assumpsit* and actions in tort were *ex delicto*. In the seventeenth century the courts began to draw procedural but not substantive distinctions between *assumpsit* and actions *ex delicto*. These distinctions became substantive differences during the nineteenth century reflecting the political, social and economic philosophy of *laissez-faire* which emphasised the importance of the legal doctrines of freedom of contract and sanctity of contract.

The area of tort which recently has been exhaustively considered by the courts is the law of negligence. The nature of this tort and the present state of the law, as the authors understand it, are dealt with in Chapter 2. **1.14**

ESSENTIALS OF A CONTRACT

Construction contracts are governed by the ordinary rules of formation and interpretation. What distinguishes construction contracts from other types of contract is their factual complexity and the widespread use of standard forms of contract. However these characteristics merely increase the burden of forensic analysis rather than changing the rules of such analysis. **1.15**

There are four essentials of a contract: **1.16**

(1) two or more parties
(2) an intention to create legal relations
(3) an agreement
(4) consideration.

Parties

There must be two or more parties present to create a contractual obligation. This statement may seem axiomatic. However the law defines party by reference to legal capacity as well as physical existence. For example if there is a parent company with subsidiary companies then contracts can be made between the parent and the subsidiary and between the various subsidiaries, provided of course they are all registered companies. However if a company operates by way of a divisional structure then the various divisions do not have a legal capacity to enter into contracts. Similarly in *Henderson* v. *Astwood* the court held that a **1.17** mortgagee who was selling a property at auction under his power of sale in the mortgage document was not entitled to bid at the auction as otherwise he would be selling to himself.

1.18   If you are drunk, insane, bankrupt, an enemy alien or a minor then your legal capacity will be impaired. The authors hasten to add that these impairments are stated as alternatives and are not cumulative!

Intention to create Legal Relations

1.19   Even though the other essentials of a contract are present the courts may consider that a promise is unenforceable if the parties are not *animo contrahendi* i.e. did not intend to create legal relations. There is a presumption in law that domestic or social arrangements are not intended to be legally binding. For example if A agrees to take B, his son, to the cinema provided that B tidies his room, there is a promisor A and consideration moving from the promisee B. Nevertheless the presumption in law will be that no legally binding obligation comes into existence. The presumption is rebuttable (i.e. need not necessarily apply) depending upon the facts of each particular case.

1.20   Clearly in commercial transactions there is no such presumption although there may be particular circumstances where the parties are not *animo contrahendi*. The House of Lords considered this point in *Independent Broadcasting Authority* v. *EMI Electronics Limited and BICC Construction Limited*. This case concerned the collapse of a cylindrical steel aerial mast at Emley Moor in Yorkshire belonging to the IBA. The mast was nearly a quarter of a mile high with a diameter at its base of only 9 feet. The mast had been designed, supplied and erected by BICC as nominated sub-contractors to EMI who were main contractors to IBA in respect of main contract works which included the aerial mast. The collapse of the mast was caused by a tension fracture of a flange in a leg of a lattice at 1027 feet. The lattice failure had been caused primarily by vortex shedding and also, to a lesser extent, by asymmetric ice loading. IBA alleged this was a design fault and proceeded against EMI for damages for breach of contract and negligence and against BICC for damages for negligence, breach of warranty and negligent mis-statement.

1.21   The claim against BICC for the breach of warranty arose from the following circumstances. The Emley Moor mast was one of a series of three cylindrical masts being designed and erected by BICC. Other masts were being erected at Belmont and Winter Hill. Work had started first at the Winter Hill site and on 16 October 1964, when the mast had reached a height of 851 feet, it began to oscillate with the result that the labour putting up a building at the foot of the mast left the site and refused to return to it. IBA raised their concerns to BICC by a letter dated 23 October when IBA expressed concern that very little was known about how such cylindrical structures were likely to behave under certain wind conditions. IBA suggested that the matter be investigated fully by taking appropriate readings from instruments attached to the Winter Hill mast in order to confirm the data upon which the design calculations had been based. BICC replied by a letter dated 11 November 1964 when they stated, *inter alia*:

'I think we will be extremely fortunate if the oscillations at Winter Hill keep within their present limits when the structure is completed. No doubt we shall learn from experience how to overcome these difficulties and I think it should be realised by all concerned that we have achieved something unique in the design of the 1250 foot mast. We expected problems with aerodynamic instability as this phenomenon is well known with cylindrical structures. *However we are well satisfied that the structures will not oscillate dangerously . . .*'

IBA alleged that the words in italics amounted to a contractual warranty. This argument was rejected by the judge at first instance but accepted by the Court of Appeal. Consideration of this particular issue was not necessary for the House of Lords' decision; however their Lordships considered the matter and disagreed with the Court of Appeal. The House of Lords rejected the finding that the assurance given in BICC's letter dated 11 November 1964 was a contractual warranty on the basis that there was no evidence that, at the time when it was given, either IBA or BICC intended the letter to create a contractual undertaking. Viscount Dilhorne stated:

'In the statement of claim it was alleged that this assurance was a warranty and that in consideration of it [IBA] forebore from requiring further investigations and from seeking independent advice as to the stability of the mast. . . . In the present case I can find nothing which can by any possibility be taken as evidence that [IBA] when [they] wrote [their] letter on [23] October or thereafter had any intention of entering into a contract or that [BICC] when [they] gave the assurance had any intention of undertaking a contractual obligation.'

It is to be noted that the representation in the IBA case was made after the **1.22** date of the contract for the construction of the Emley Moor television mast. That is to say the representation was not collateral to the main contract; the contractual result may have been different if the representation had been made before or at the time of the Emley Moor contract and the court may have found that there was a collateral contract.

1.23   By way of contrast to the decision in *IBA* is the case of *Edwards* v. *Skyways Limited*. E was an aircraft pilot employed by S and a member of S's contributory pension fund which entitled him on leaving the defendants' employment in advance of retirement age to a choice of either withdrawing the sum of his own contributions to the fund or taking the right to a paid-up pension payable at retirement age. In January 1962 some 15% of S's pilots, including E, were made redundant. The pilots' union BALPA agreed with S's account (as recorded in the notes of the meeting) that:

> 'pilots declared redundant and leaving [the defendant company] would be given an *ex gratia* payment equivalent to the [defendant company's] contributions to the pension fund.'

On the evidence at trial it appeared that this was an incorrect record of what was agreed as S's representative had said at the meeting that the company would make *ex gratia* payments 'approximating to' S's contribution. E elected to withdraw his contributions to the pension fund and to receive the *ex gratia* payment. S paid the contribution but refused to pay the *ex gratia* payment contending that the recorded agreement was not intended to create legal relations and was too vague and thus was not legally binding. The court held that where there was an agreement, the subject of which related to business affairs, the onus of establishing that the agreement was not intended to create legal relations, which was on the party setting up that defence, was a heavy onus. S had failed to discharge this burden for the following reasons, namely the words *ex gratia* were used simply to indicate that the party agreeing did not admit any pre-existing liability on S's part, and the mere use of the phrase *ex gratia* as part of a promise to pay did not show that the promise when accepted should have no binding effect in law. Further the use of the words *approximating to* did not render the terms of the agreement too vague to be enforceable for at most the phrase would denote on the evidence a rounding off of a few pounds downwards to a round figure.

The *Edwards* case has recently been considered by the Court of Appeal in *Kleinwort Benson Ltd* v. *Malaysia Mining Corporation Bhd* when the court found that the presumption in *Edwards* was not appropriate in deciding whether or not the relevant documents contained a contractual promise. In *Kleinwort* the court held that a letter of comfort from a parent company to a lender stating that it was the policy of the parent company to ensure that its subsidiary was 'at all times in a position to meet its liabilities' in respect of a loan made by the lender to the subsidiary, did not have contractual effect if it was merely a statement of present fact regarding the parent company's intentions and was not a contractual promise as to future conduct.

Agreement

The presence or otherwise of an agreement is determined by an objective **1.24** test and not a subjective test. Objectivity is based upon a reasonable man's understanding of a particular set of circumstances or facts. In its purest form the test excludes that which was actually in the minds of the parties although as a matter of practice when the courts are exploring the issue of concensus they sometimes blur the distinction between objectivity and subjectivity. Of the objective test in relation to an intention to create legal relations Megaw J in the *Edwards* case stated:

'I am not sure that I know what that means in this context. I do however think that there are grave difficulties in trying to apply a test as to the actual intention or understanding or knowledge of the parties especially where the alleged agreement is arrived at between a limited liability company and a trade association; and especially where it is arrived at at a meeting attended by five or six representatives on each side. Whose knowledge, understanding or intention is relevant? But if it be the "objective" test of the reasonable man, what background knowledge is to be imputed to the reasonable man, when the background knowledge of the ten or twelve persons who took part in arriving at the decision no doubt varied greatly between one another?'

An exhaustive analysis of case law to establish the borderline between **1.25** objectivity and subjectivity goes well beyond the ambit of this book. Suffice it to say that, whilst the courts have well in mind the basic rules as to formation of contract, they sometimes adopt a pragmatic solution to the forensic problems created by an application of the rules. That being said, it is a basic rule that agreement is evinced by offer and acceptance.

The Offer

1.26 An offer is a written or oral statement by a person of his willingness to enter into a contract upon terms which are certain or are capable of being made certain. The offeror's intention to enter into a contract may be actual or apparent. An apparent intention is determined by the objective test discussed above.

1.27 An offer must be distinguished from what is merely a request for information or an invitation to treat. An invitation to treat is a request for an offer. Neither a request for information nor an invitation to treat can be converted into a binding contract by an acceptance. Competitive tendering is a common method of procurement in the construction industry. In the absence of special circumstances the invitation to tender sent out to contractors by the employer, or by the consultant on behalf of the employer, is an invitation to treat and not an offer. It is the contractor's submission of tender which constitutes the offer and which in turn must be accepted by the employer to give rise to a formally binding contract. For this reason the employer is not bound by the lowest tender nor as a general rule is the employer responsible for the contractor's costs of preparing the tender.

1.28 An offer can be withdrawn at any time prior to it being accepted. This is so even though the offeror stipulates that the offer shall be kept open for a particular period of time. In *Routledge* v. *Grant* G offered to buy a house from R stating that R had six weeks in which to decide whether or not to sell. Before the expiration of the six week period G withdrew his offer and the court held that he was entitled to do this. In order for the offeror to be bound by a statement that an offer shall be open for acceptance for a particular period that statement must form part of a contract of option which is a contract having a separate legal existence to the main transaction.

1.29 The withdrawal of an offer is not effective until the withdrawal has been communicated to the offeree. This communication may be written or oral. If as in *Routledge* v. *Grant* the offeror has stated that the offer will be open for a specified time then the offer will lapse at the end of the relevant period even though there has not been any communication of withdrawal between offeror and offeree. It is also arguable that where no time is specified an offer may lapse after a reasonable period of time and no longer be capable of being accepted even though no communication of an intention to withdraw the offer has been made to the offeree. What is a reasonable time will depend upon the particular circumstances of each case.

The Acceptance

An acceptance is a written or verbal expression of assent to the terms of **1.30** the offer. It must be unequivocal and unconditional. For example an acknowledgement of the *receipt* of an offer is not an unequivocal acceptance nor is an acceptance which fails to distinguish between alternative offers. In *Peter Lind & Co. Limited* v. *Mersey Docks and Harbour Board* the employer, the Mersey Docks and Harbour Board, invited the building contractor to submit alternative tenders for the construction of a container freight terminal, one tender to be at a fixed price and the other at a price varying with the cost of labour and materials. The employer wrote to the contractor stating that they accepted 'your tender'. The court held that this letter was imprecise because it did not specify which tender was being accepted so that there was no concluded contract at that stage.

A conditional acceptance is not effective; indeed a conditional **1.31** acceptance may on its wording be a counter offer terminating the original offer so that it is no longer capable of being accepted even though an unconditional acceptance is subsequently sent to the offeror. In *Hyde* v. *Wrench* W offered to sell his farm to H for a price of £1000. H made a counter offer in the sum of £950 which was rejected. H then purported to accept the original offer. However the court held he was no longer entitled to do so and there was no contract for the sale of the farm.

There is a general rule that an acceptance must be communicated by **1.32** the offeree to the offeror to create a binding contract. In the absence of any particular requirements set out in the offer, communication of an acceptance may be oral or in writing. There are certain exceptions to the general rule. *Firstly* posted acceptances are deemed to take effect from the date of posting regardless of whether or not the letter actually reaches the offeror. The same rule applies to telegrams but not to instantaneous communications such as telephone, telex or facsimile transmission. *Secondly* communication of acceptance is not necessary for the creation of a unilateral contract as compared to a synallagmatic, or bilateral, contract. A synallagmatic contract involves the mutual exchange of promises whereas a unilateral contract binds only one party — for example an offer to reward a person who supplies information in relation to a lost article. In *Carlill* v. *The Carbolic Smoke Ball Company* the defendants who were proprietors of a medical preparation called 'the Carbolic Smoke Ball' issued an advertisement in which they offered to pay £100 to any person who contracted influenza after having used one of their smoke balls in a specified manner for a specified period. C purchased one of the smoke balls on the faith of the advertisement, used it in the specified manner and subsequently contracted influenza. The Carbolic Smoke Ball Company refused to pay over the £100 and upon being sued for breach of contract the court held that C was entitled to recover. Bowen LJ stated:

'It is not a contract made with all the world. . . . It is an offer made to all the world; and why should not an offer be made to all the world which is to ripen into a contract with anybody who comes forward and performs the condition.'

Thirdly the offerees' conduct may preclude him from relying upon a failure of communication of an acceptance. Denning LJ considered this problem in *Entores Limited* v. *Miles Far Eastern Corporation* when he stated:

'Now take a case where two people make a contract by telephone. Suppose for instance, that I make an offer to a man by telephone and, in the middle of his reply, the line goes "dead" so that I do not hear his words of acceptance. There is no contract at that moment. The other man may not know the precise moment when the line failed. He will know that the telephone conversation was abruptly broken off, because people usually say something to signify the end of the conversation. If he wishes to make a contract, he must therefore get through again so as to make sure that I heard. Suppose next that the line does not go dead, but it is nevertheless so indistinct that I do not catch what he says and I ask him to repeat it. He repeats it and I hear his acceptance. The contract is made, not on the first time when I do not hear, but only the second time when I do hear. If he does not repeat it, there is no contract. The contract is only complete when I have his answer, accepting the offer.'

1.33 Silence cannot be taken as an acceptance of an offer even though the offer stipulates the same. This is to be contrasted with conduct which may constitute a binding acceptance. Consider for example the situation where a sub-contractor for the supply of ready mixed concrete has forwarded an offer in response to a main contractor's enquiry. The main contractor responds with a purchase order which sets out the main contractor's standard terms of contract some of which are at variance with the terms of the sub-contractor's offer. The sub-contractor intends to negotiate these matters with the main contractor, however unbeknown to the sales department of the sub-contractor the first deliveries of concrete have been sent from the batching plant. The next day the sub-contractor communicates with the main contractor notifying the main contractor that its terms and conditions of the contract are not acceptable. In such circumstances the courts have held that the delivery of the concrete by the sub-contractor to the main contractor constitutes an acceptance, by conduct, of a contract based upon the terms of the main contractor's purchase order.

The 'Last Shot' Doctrine

Commercial negotiations, particularly in the construction industry, can **1.34** be both protracted and tortuous; offers being met with counter offers and counter offers with counter counter offers, each party endeavouring to impose a standard form of contract or their own terms and conditions of trade. As stated above the law of contract requires the offer and the acceptance to be in the same terms, and the courts are often faced with a difficult burden of forensic analysis to decide whether, and if so what, contract came into existence. The 'last shot' doctrine provides that where there is a series of conflicting documents they shall all be treated as counter offers and the contract, if indeed one comes into existence at all, is to be based upon the last document in time. Whilst this is useful doctrine for assisting the forensic process it is not an absolute doctrine and each contract must be considered on its own particular facts. Consider for example the facts of *OTM Limited* v. *Hydranautics*. On 8 September 1978 H offered to sell to OTM a device for tensioning chains on a monitoring buoy in the North Sea. The offer incorporated H's terms and conditions including a Californian arbitration clause and a Californian proper law clause i.e. that Californian law should govern the formation, construction and performance of the contract. On 29 September OTM telexed H stating:

'It is our intention to place an order for one chain tensioner. . . . A purchase order will be prepared in the near future but you are directed to proceed with the tensioner fabrication on the basis of this telex. The purchase order will be issued subject to our usual terms and conditions.'

Three days later OTM sent to H a purchase order which had the following condition:

> 'Acceptance of contract: The written acceptance of this contract, the commencement of performance pursuant thereto . . . by the sellers constitutes an unqualified acceptance by the seller of all the terms and conditions of this contract. This contract . . . constitutes the entire agreement between the parties either oral or written . . .'

The purchase order led to an exchange of telexes. However H made no objection to the above condition although they complained that they had commenced work on the understanding that their offer was acceptable and they were now facing the introduction of new contractual terms. Negotiations continued and an agreement was reached on variations to the terms of the purchase order. On 20 October OTM sent a telex to H agreeing to the one outstanding point and asked H whether in view of the changes to the purchase order H would prefer that OTM re-issued their purchase order. On 20 December H replied to the effect that they saw no need to re-issue the purchase order and enclosed a formal acknowledgement of order which contained the following clause:

> 'Acceptance of buyers' order is conditional and subject to . . . the following conditions. . . . Unless buyer shall notify seller in writing to the contrary within five days of receipt of this document the buyer shall be deemed conclusively to have accepted the exact terms and conditions hereof.'

A copy of this acknowledgement of order was signed by OTM and returned to H on 3 January. The court held that OTM's telex dated 29 September was not an acceptance of H's offer dated 8 September, it was a letter of intent. Further OTM's purchase order dated 5 October was a counter offer which destroyed the original offer in total. The contract was concluded by OTM's telex sent on 20 October. The clause in H's letter dated 20 December was meaningless since there was nothing left to accept; the contract had already been made.

Incomplete Agreements
1.36 To create a legally binding contract the process of offer and acceptance must result in the parties being in agreement on all the terms which are essential to their bargain; there must be a *consensus ad idem*.

As a general rule the essential terms of a construction contract are **1.37** parties, description of the works, price and period for construction. A failure to agree on price or time for performance is not necessarily fatal; in certain circumstances the courts may supply terms as to reasonable price and a reasonable period for performance. Such terms are now implied by the Supply of Goods and Services Act 1982, sections 14 and 15.

A particular difficulty arises in respect of collateral warranties in so far **1.38** as the ultimate beneficiary of the warranty might not be known at the time the parties enter into the principal contracts. Does this constitute a failure to agree an essential term?

Lord Buckmaster stated in the case of *May and Butcher Ltd* v. *The King* **1.39** that:

'It has long been a well recognised principle of contract law that an agreement between two parties to enter into an agreement in which some critical part of the contract matter is left undetermined is no contract at all. It is of course perfectly possible for two people to contract that they will sign a document which contains all the relevant terms, but it is not open to them to agree that they will in the future agree upon a matter which is vital to the arrangement between them and has not yet been determined.'

Viscount Dunedin added:

'To be a good contract there must be a concluded bargain and a concluded contract is one which settles everything that is necessary to be settled and leaves nothing to be settled by agreement between the parties. *Of course it may leave something which still has to be determined but then that determination must be a determination which does not depend upon the agreement between the parties. . . .* Therefore you may very well agree that a certain part of the contract of sale, such as price, may be settled by someone else. . . . As long as you have something certain it does not matter. For instance with regard to price it is a perfectly good contract to say that the price is to be settled by the buyer . . .'

Paragraphs 6.34 and 6.35 below set out examples of precedents which **1.40** have been used to try and overcome the problem of a non-existent party. The precedent set out in paragraph 6.34 clearly runs a risk of being nothing more than an agreement to agree and therefore unenforceable in law. However in contrast, whilst the precedent set out in paragraph 6.35 still leaves undecided the identity of the party who is to have the benefit of the collateral warranty, that omission can be dealt with by the *determination* of the developer, that is to say applying the words of Viscount Dunedin: 'A determination which does not depend upon the agreement between the parties.'

1.41 The case of *Foley* v. *Classique Coaches Ltd* provides a useful illustration of how a contract can be rescued by the use of an implied term. This case concerned an agreement by F to sell to C certain land which C intended to use for their business of motor coach operators. The sale was subject to a second agreement in which C agreed to purchase from F all the petrol required for the purposes of their business at 'a price to be agreed by the parties in writing and from time to time'. The land was conveyed to C and the petrol agreement operated for a period of three years. Thereafter disputes arose and C purported to repudiate the petrol agreement alleging that it was not binding because there was no agreement as to the price of the petrol. The court rejected this argument holding that a term must be implied in the agreement that the petrol supplied by F should be of a reasonable quantity and sold at a reasonable price and that if any dispute arose as to what was a reasonable price it was to be determined by an arbitration clause in the petrol agreement. The agreement was therefore valid and binding.

Consideration

1.42 Save for contracts made under seal, the courts will not enforce gratuitous promises. There must be valuable consideration. Valuable consideration is 'something of value in the eye of the law': *Thomas* v. *Thomas*. Clearly the payment of money or a promise to pay money is valuable consideration. However 'in the eyes of the law' other acts, however insignificant, may provide valuable consideration. For example a promise to give £50 'if you will come to my house' was held to be valuable consideration in *Gilbert* v. *Ruddeard*. However, as a general rule, a moral obligation does not provide valuable consideration, for example a promise made 'in consideration of natural love and affection' *Bret* v. *J.S.* Nor is a pre-existing legal obligation sufficient to provide valuable consideration, often referred to as 'past consideration'. The giving of a collateral warranty by an architect in consideration of terms of appointment that have already been fulfilled or the giving of a guarantee by a construction company in consideration of a payment under the construction contract, are examples of past consideration, cf. the decision in *Williams* v. *Roffey Bros. & Nicholls (Contractors) Ltd.*

Valuable consideration need not be adequate consideration in a sense **1.43** that the courts are not concerned as to the fairness of the bargain between two contracting parties. For this reason payment of nominal consideration is sufficient. Collateral warranty agreements usually provide that consideration for the agreement shall be the payment by the promisee to the promisor of the sum of £1 or £10. Whether it be £1 or £10, this consideration, though nominal, is valuable consideration for the purposes of enforcing the warranty.

Valuable consideration must support the promise and not the contract. **1.44** That is to say there must be some detriment to the promisee (the giving of value) or some benefit to the promisor (receipt of value). In many cases a detriment from the promisee will be the same as a benefit to the promisor: for example a sub-contractor promises to carry out piling works in consideration for which the main contractor promises to pay the price of those works. The carrying out of the piling works and the promise to pay the price of the works are respectively both detriments and benefits. However whilst consideration must move from the promisee, that is to say the promisee must suffer some detriment, it is not necessary for consideration to move to the promisor, that is to say the promisor need not obtain a benefit under the contract. For example if A promises to B that A will guarantee the repayment of a loan of finance to be made by B to C, whilst A derives no benefit from the transaction, if B, in consideration of A's promise, makes the loan to C, then B will suffer a detriment which provides valuable consideration moving from B, the promisee, to A, the promisor, in respect of a contract of guarantee. Similarly in the *Shanklin Pier* case referred to above, if Shanklin had not procured that the contractor Carter entered into a contract with Detel for the use of their paints, there would not have been valuable consideration moving from Shanklin, the promisee, to Detel, the promisor, whereby Detel's representations as to the quality of their paint were a contractual warranty.

FORM OF CONTRACT

Unless the contract is a deed there are no special rules governing the form **1.45** of a construction contract. The contract may be a written contract or an oral contract or partly written and partly oral.

Deeds

Much of the law relating to the execution of deeds was swept away on 31 **1.46**

July 1990, when the Law of Property (Miscellaneous Provisions) Act 1989, dealing with the execution of deeds by individuals, and the Companies Act 1989 dealing with the execution of deeds by companies both came into operation.

1.47   Section 1(3) of the Law of Property (Miscellaneous Provisions) Act provides that a deed must be signed by the person who is to be bound and the signature must be attested by a witness. If this procedure is not followed then the document will not be valid as a deed, but it will be valid as a simple contract if there is valuable consideration. Further section 1(2)(a) provides that a document shall not be a deed unless it is made clear on its face that it is intended to be a deed by the person making it. It follows that the document must be described as a deed and the attestation clause must include a statement that it is being executed as a deed. If such descriptive words are not included then the document will not be valid as a deed. The Act does not abolish the old common law rule that to be effective a deed must be delivered by the person who is to be bound. Delivery does not mean the physical handing over of the document but evincing an intention, whether by words or conduct, that the party is to be bound. Such intention was satisfied prior to the Act by the act of signature and it is submitted that signature will continue to satisfy the requirement of delivery.

1.48   Companies may still use their seals to execute a deed provided that this is in accordance with their articles of association. Section 130 of the Companies Act 1989, introducing a new section 36A to the Companies Act 1985, provides by sub-section 4 that if a document is signed by a director and the secretary of the company or by two directors of the company and is expressed (in whatever form of words) to be executed by the company, it will have the same effect as if it had been executed under the common seal of the company i.e. a deed. Further sub-section 5 provides that a document which makes it clear on its wording that it is intended to be a deed has the effect of a deed even though the specific requirements for execution dealt with in sub-section 4 above are not complied with. For example a document which is described as this deed and is executed by one director of the company will operate as a deed. Extreme care must be taken to ensure that if it is intended only to create a simple contract, a company giving a collateral warranty does not accidentally execute the document in such a manner as to give rise to a deed.

Consideration and Limitation Periods

1.49   There are two important differences between contracts which are not deeds (referred to as 'simple contracts') and contracts which are under seal. *First* simple contracts and deeds have different limitation periods. See paragraph 5.52 below. *Secondly* unlike a simple contract, a deed does not have to be supported by valuable consideration. Thus where a collateral warranty consists of unilateral undertakings by one party the contract must be a deed if it is to be enforceable. It is important to note that, whilst consideration is not necessary for a deed, in the absence of valuable consideration and, arguably, in the absence of something more than mere nominal consideration, the remedy of *specific performance* will not be available in respect of the contractual undertakings: *Milroy* v. *Lord*. Specific performance is an equitable remedy requiring the contract breaker to fulfil his contractual obligations rather than awarding damages for breach. Equity does not assist a volunteer hence the need for consideration. The authors suggest that a deed which relies on merely nominal consideration could be rescued by the addition of a consideration which consists of 'the mutual undertakings' of the parties to the deed.

CONSTRUING A CONTRACT

The same rules of construction apply to both simple contracts and to deeds. **1.50**

The general rule is that whilst the court strives to give effect to the intention of the parties, it must give effect to that intention as expressed by the words used in the agreement. That is to say it must ascertain the meaning of the words actually used and not try to interpret the motive or state of mind of the parties: *Inland Revenue Commissioners* v. *Ravielle* and *Prenn* v. *Simmonds*. **1.51**

The courts must have regard to the ordinary meaning of words used unless they are technical or scientific. That being said the courts are concerned to ascertain the intention of the parties and not merely indulge in semantic exercises. In *Lloyd* v. *Lloyd* Lord Cottenham stated: **1.52**

'If the provisions are clearly expressed, and there is nothing to enable the court to put upon them a construction different from that which the words import, no doubt the words must prevail; but if the provisions and expressions be contradictory and if there be grounds, appearing on the face of the instrument, affording proof of the real intention of the parties, then that intention will prevail against the obvious and ordinary meaning of the words. If the parties have themselves furnished a key to the meaning of the word used, it is not material by what expression they convey their intention'.

1.53 In construing particular terms of a contract the whole of the contract must be considered. Further if there is no ambiguity or uncertainty then the court must give effect to the intention of the parties however harsh. In *Trollope & Colls Ltd* v. *North West Metropolitan Regional Hospital Board* the contract provided for three phases of work, phase 1 to be completed by 30 April 1969 and phase 3 to commence six months after the date of practical completion of phase 1 and to be completed on 30 April 1972 i.e. a construction period of 30 months for phase 3. The completion of phase 1 was delayed by a period of 59 weeks, 47 of which were not the fault of the contractor. Practical completion of phase 1 was not achieved until 22 June 1972, leaving a construction period of only 16 months, rather than 30 months, for the completion of phase 3. The court took the view that the intention of the parties was perfectly clear from the wording of the contract and refused to relax the contractors' obligations by implication of the term that the time for completion of phase 3 should be extended by the same period as the extention of time granted in respect of phase 1.

1.54 Where the parties have made manuscript deletions, additions or amendments to a printed form in the event of any ambiguity in construing the document as a whole the manuscript will be given greater weight than the printed terms: *Sutro & Co.* v. *Heilbut Symons and Co.* Where there is inconsistency between figures and words the court will have regard to the words before the figures: *Saunderson* v. *Piper*.

1.55 If there is inconsistency between different parts of the same contract then, in the absence of an express term resolving the ambiguity (for example condition 2.2 and 2.4 of JCT 'With Contractor's Design' 1981 Edition), the courts will endeavour to give effect to that part of the contract which expresses the real intention of the parties: *Walker* v. *Giles*.

1.56 If the contract document or documents establish a clear intention the courts adopt an interventionist approach to resolve any difficulties flowing from the actual words used by the parties: *Gwyn* v. *Neath Canal Co.*:

'The result of the authorities is that, when a court of law can clearly collect from the language within the four corners of a deed or instrument in writing the real intention of the parties, they are bound to give effect to it by supplying anything necessary to be inferred from the terms used, and by rejecting as superfluous whatever is repugnant to the intention so discerned.'

For example in *Mourmand* v. *Le Clair* the parties expressed the repayment   **1.57**
of a debt to be by instalments of 'seven' on a particular day of the month;
the court inserted the word 'pounds' after the word 'seven'. In *Simpson* v.
Vaughan a debtor gave an acknowledgment of debt which was stated to
be 'for money borrowed which I promise *never* to pay'; the court struck
out the word 'never'.

Parol Evidence Rule
Where it is clear that the parties intended the whole of their agreement to   **1.58**
be set out in one document or a series of documents the operation of the
parol evidence rule will prevent the parties from relying on extrinsic
evidence to try and amend the terms of their agreement. For example
evidence of negotiations that took place prior to the conclusion of the
contract will not be admissible.

The parol evidence rule applies to both oral evidence and also to   **1.59**
documentary evidence, for example letters and minutes of meetings. It is
not an absolute rule however, and there are exceptions as the courts adopt
a commonsense approach to the question of interpretation of contracts.
In particular if there is some uncertainty or ambiguity as to the intentions
of the parties then the courts are prepared to look at what has been
described as the *matrix* of a contract, that is to say the surrounding
circumstances, background and commercial purpose of the agreement:
Prenn v. *Simmonds*.

Recitals
Recitals are the introductory statements in a written agreement or deed   **1.60**
setting out a précis of the parties' intentions. Recitals usually appear in
documents after the words 'whereas' and before the words 'now it is
hereby agreed as follows', the latter phrase introducing the operative or
main conditions of the agreement. If there is any ambiguity or uncertainty
arising from the operative or main conditions of the agreement the courts
will look at the recitals in order to establish the intention of the parties to
the agreement. It is also important to note that the intention of the parties
as evinced by the recitals may be relevant to the court's consideration of
whether or not to imply a term into the agreement. Implied terms are
dealt with in paragraph 1.65 below.

Ejusdem Generis Rule
This is a rule of construction which provides that, where a contract   **1.61**
condition or clause sets out a list of specific matters so as to create a
common category and the specific matters are followed by general words,
the courts will construe the general words restrictively so as to confine

those words to the common category. For example the phrase 'and all other deleterious materials' coming after a list of deleterious materials will be construed restrictively.

Contra Proferentem

1.62 This is another rule of construction applying to written documents or deeds. The rule provides that if the wording of an agreement is ambiguous or uncertain, *but not otherwise*, the contract should be construed more strongly against the person whose words they are rather than the other party:

> 'We are presented with two alternative readings of this document and the reading which one should adopt is to be determined, amongst other things, by a consideration of the fact that the defendants put forward the document. They have put forward a clause which is by no means free from obscurity and have contended . . . that it has a remarkably, if not extravagantly, wide scope and I think that the rule *contra proferentem* should be applied': *John Lee & Son (Grantham) Limited* v. *Railway Executive*.

1.63 If a party has incorporated its own standard terms and conditions of trade into an agreement then in the event of ambiguity those terms and conditions will be construed *contra proferentem* that party. Where however the parties execute standard forms of contracts such as JCT 80, the *contra proferentem* rule will only operate in respect of amendments or additions to the contract.

1.64 *Contra proferentem* is a particularly important rule of construction in relation to exclusion clauses. Exclusion clauses are dealt with in paragraph 8.72 below.

IMPLIED TERMS

1.65 The express terms of a contract do not necessarily constitute all the relevant terms of the agreement. In certain circumstances the courts are prepared to imply terms into a contract provided such terms are necessary to give *business efficacy* to the agreement. The leading case on implied terms is *The Moorcock*. In that case Bowen LJ stated:

> 'Now, an implied warranty, or, as it is called, a covenant in law, as distinguished from an express contract or express warranty, really is in

all cases founded upon the presumed intention of the parties, and upon reason. The implication which the law draws from what must obviously have been the intention of the parties, the law draws with the object of giving efficacy to the transaction and preventing such a failure of consideration as cannot have been within the contemplation of either side; and I believe if one were to take all the cases, and there are many, of implied warranties or covenants in law, it will be found that in all of them the law is raising an implication from the presumed intention of the parties with the object of giving the transaction such efficacy as both parties must have intended that at all events it should have.'

Terms may be implied as a matter of law. That is to say they are implied **1.66** as a matter of policy and are of general application to all contracts. This matter is dealt with in detail in Chapter 4. Further terms may be implied as a matter of fact. That is to say as a matter of construction of the presumed intention of the parties to a particular contract. The *Foley* case referred to in paragraph 1.41 is an illustration of a term being implied as a matter of fact.

Whilst a term will not be implied unless in the particular circumstances **1.67** of each case it is reasonable to imply such a term, this does not mean that a term will be implied merely because it is reasonable. For example see the *Trollope & Colls* case referred to in paragraph 1.53 above where the court refused to imply a term to render a harsh contract more reasonable. Further a term will not be implied if it is inconsistent with the express terms of the contract. In *Martin Grant & Co. Limited* v. *Sir Lindsay Parkinson & Co. Limited* the Court of Appeal refused to imply a term in a building sub-contract:

'that (a) the [main contractors] would make sufficient work available to the [sub-contractors] to enable them to maintain reasonable progress and to execute their work in an efficient and economic manner; and (b) the main contractors should not hinder or prevent [the sub-contractors] in the execution of the sub-contract works'

where the express conditions of the sub-contract provided for a 'beck and call' obligation on the sub-contractor. That is to say the sub-contractor was obliged to carry out his works '. . . at such time or times and in such manner as the [main contractor] shall direct or require'.

Chapter 2

The Rise of Collateral Warranties

2.1   In order to understand why collateral warranties have assumed such importance, it is necessary to look at the tort of negligence in relation to construction problems; this necessarily involves some consideration of the law of negligence and its development immediately prior to the House of Lords decision in 1988 in *D & F Estates* v. *The Church Commissioners for England*, the *D & F* case itself and the subsequent decision in the House of Lords in 1990 in *Murphy* v. *Brentwood District Council*.

NEGLIGENCE

2.2   The tort of negligence is concerned with breach of a duty to take care. In order to succeed in an action for negligence, a plaintiff must prove:

(1) the defendant owed to the plaintiff a legal duty of care; and
(2) the defendant was in breach of that duty; and
(3) the plaintiff has suffered damage as a result of that breach.

2.3   The legal duty of care referred to is one that arises independently of a contractual obligation and, indeed, in the absence of contract. Over many years, the courts have produced a long series of decisions to assist in deciding whether or not, on particular facts, a duty of care arises.

2.4   The modern law of negligence really begins in 1932 when the famous decision in *Donoghue* v. *Stevenson* reached the House of Lords. A young lady was bought a bottle of ginger beer by a friend. She had drunk some of the ginger beer, which was in an opaque bottle, before she discovered that there was a decomposing snail in the bottle. It was alleged that she became ill as a result. There was no question in this case of the friend bringing an action in contract under the Sale of Goods Act against the retailer from whom the ginger beer had been purchased because the friend had not suffered any damage. The young lady could not sue the retailer because she had no contract herself with him.

It was in this way that the House of Lords came to be asked whether   **2.5** the young lady had a cause of action in negligence against the manufacturer. They held by a majority that a manufacturer who sold products in such a form that they were likely to reach the ultimate consumer in the state in which they left the manufacturer with no possibility of intermediate examination owed a duty to the consumer to take reasonable care to prevent injury. Some understanding of the radical development in English law that this case represented can be gained from the dissenting judgment of Lord Buckmaster who did not agree with the majority in the House of Lords:

> 'There can be no special duty attaching to the manufacturer of food apart from that implied by contract or imposed by statute. If such a duty exists, it seems to me it must cover the construction of every article, and I cannot see any reason why it should not apply to the construction of a house. If one step then why not 50? If a house be, as it sometimes is, negligently built and in consequence of that negligence the ceiling falls and injures the occupier or anyone else, no action against the builder exists according to the English law, although I believe such a right did exist according to the laws of Babylon.'

Little did Lord Buckmaster, in his dissenting judgment, appreciate how   **2.6** the *Donoghue* case would be the basis for a rapid expansion of the law of tort in negligence over the following 56 years, along the very lines that he robustly refused to contemplate in his judgment. However, his view did not prevail in 1932; in the same case, Lord Atkin formulated a principle so as to test whether a duty of care exists:

> 'The liability for negligence whether you style it such or treat it, as in other systems, a species of *culpa,* is no doubt based upon a general public sentiment of moral wrong doing for which the offender must pay. But acts or omissions which any moral code would censor cannot in a practical world be treated so as to give a right to every person injured by them to demand relief. In this way, rules of law arise which limit the range of complaints and the extent of their remedy. The rule that you are to love your neighbour becomes in law, you must not injure your neighbour; and the lawyer's question "who then is my neighbour?" receives a restrictive reply. You must take reasonable care to avoid acts or omissions which you can reasonably foresee would be

likely to injure your neighbour. Who, in law, is my neighbour? The answer seems to be — persons who are so closely and directly affected by my act that I ought reasonably to have them in contemplation as being so affected when I am directing my mind to the acts or omissions which are called into question.'

2.7   The law in this respect remained fairly static until the early 1960s; through the 1960s and 1970s there was rapid development of the law of negligence, in particular in construction cases.

1932 TO 1988

2.8   It took from *Donoghue* in 1932 until 1964 to extend the principle of the *Donoghue* decision to statements which were given negligently: *Hedley Byrne & Co. Limited* v. *Heller & Partners Limited.* Advertising agents, Hedley Byrne, needed a reference from a banker as to the creditworthiness of a potential customer. They approached their bankers who sought the advice of merchant bankers who in turn reported to Hedley Byrne. The report was headed 'without responsibility' and said that the potential customer was good for ordinary business arrangements. Hedley Byrne proceeded with their contract and by reason of the customer not being good for ordinary business arrangements, lost a considerable sum of money. They sued the merchant bankers. The House of Lords held that a person is liable for statements made negligently in circumstances where he knows that those statements are going to be acted on and they were acted on. However, in this case, the merchant banker escaped liability by reason of having expressed their report to be without responsibility. This case may have assumed new importance since the decision in *Murphy* v. *Brentwood District Council* (see 2.23).

2.9   The first major extension of the test of Lord Atkin in *Donoghue* in a building case was in 1972 in *Dutton* v. *Bognor Regis Urban District Council and Another* (now overruled by *Murphy*, see 2.24). A house was built on a rubbish tip and D was the second owner of the house. The walls and the ceiling cracked, the staircase slipped and the doors and windows would not close; the damage was caused by inadequate foundations. D sued the builder (with whom she settled before the hearing) and the local authority. The Court of Appeal held that the local authority, through their building inspector, owed a duty of care to D to ensure that the inspection of the foundations of the house was properly carried out and that the foundations were adequate, and the local authority were liable to

D for the damage caused by the breach of duty of their building inspector in failing to carry out a proper inspection of the foundations. Lord Denning MR said, applying Lord Atkin's test:

> 'I should have thought that the inspector ought to have had subsequent purchasers in mind when he was inspecting the foundations — he ought to have realised that, if he was negligent, they might suffer damage.'

The *Dutton* case was followed on this point in many subsequent and important cases: *Sparham-Souter* v. *Town & Country Developments (Essex) Limited*; *Sutherland & Sutherland* v. *C R Maton & Son Limited*; *Anns* v. *Merton London Borough Council*; *Batty and Another* v. *Metropolitan Property Realisations Limited and Others*. However, as is often the case with changes in orthodoxy in the law, the seeds of destruction of this widening of the law of negligence were sown by a passage, albeit *obiter dicta*, in the judgment, that of Stamp LJ in the *Dutton* case in 1972, but it was to take many more years before change came:

2.10

> 'I may be liable to one who purchases in the market a bottle of ginger beer which I have carelessly manufactured and which is dangerous and causes injury to personal property; but it is not the law that I am liable to him for the loss he suffers because what is found inside the bottle and for which he has paid money is not ginger beer but water. I do not warrant, except to an immediate purchaser, and then by contract and not in tort, that the thing I manufacture is reasonably fit for its purpose. The submission is, I think, a formidable one and in my view raises the most difficult point for decision in this case. Nor can I see any valid distinction between the cases of a builder who carelessly builds a house which, although not a source of danger to personal property, nevertheless, owing to a concealed defect in its foundations, starts to settle and crack and becomes valueless, and the case of a manufacturer who carelessly manufactures an article which, though not a source of danger to a subsequent owner or to his other property, nevertheless owing to hidden defect quickly disintegrates. To hold that either the builder or the manufacturer was liable except in contract would be to open up a new field of liability the extent of which could not, I think, be logically controlled and since it is not in my judgment necessary to do so for the purposes of this case, I do not more particularly, because of the absence of the builder, express an opinion whether the builder has a higher or lower duty than the manufacturer.'

2.11   In 1978, the position altered again with *Anns* v. *Merton London Borough Council*. Lessees of flats claimed against a local authority in negligence in relation to the local authority's powers of inspection under the by-laws in that, it was said, they had allowed the contractors to build foundations in breach of the by-laws, with resulting damage to the flats. In a passage in his speech, which later became the excuse for a far reaching and dramatic expansion of the circumstances in which a duty of care might be held to exist, Lord Wilberforce said in the House of Lords:

> 'Through the trilogy of cases in this House, *Donoghue* v. *Stevenson*, *Hedley Byrne & Co. Limited* v. *Heller & Partners Limited* and *Home Office* v. *Dorset Yacht Co. Limited*, the position has now been reached that in order to establish that a duty of care arises in a particular situation it is not necessary to bring the facts of that situation within those of previous situations in which a duty of care has been held to exist. Rather the question has to be approached in two stages. First one has to ask whether as between the alleged wrong-doer and the person who has suffered the damage there is a sufficient relationship of proximity or neighbourhood such that, in the reasonable contemplation of the former, carelessness on his part may be likely to cause damage to the latter in which case a *prima facie* duty of care arises. Secondly, if the first question is answered affirmatively, it is necessary to consider whether there are any considerations which ought to negative, or to reduce or limit the scope of the duty or class of person to whom it is owed, or the damage to which a breach of it may give rise.'

2.12   The development and extension of the law of tort probably reached its climax in the House of Lords in *Junior Books Limited* v. *Veitchi Co. Limited* in 1983. That case was on appeal to the House of Lords from Scotland. Specialist flooring sub-contractors had laid a floor at the employers' factory and it was said by the employers that the floor was defective. The employers, who were not in contract with the sub-contractor, brought an action in *delict* (which is substantially the same cause of action in Scotland as negligence is in England). Despite the absence of any allegation by the employer that there was a present or imminent danger to the occupier (an essential ingredient of the *Anns* decision), the employers succeeded in their argument that the sub-contractor owed them a duty of care in negligence and that the sub-contractor was in breach of that duty. It was said that there was a close commercial relationship between the employers and the sub-contractors. It is that justification for the decision which had led to enormous difficulties in the minds of those closely involved with the construction industry — after all, what is unusual about an employer engaging a contractor who in turn engages sub-contractors? Notwithstanding these difficulties, Goff LJ in *Muirhead* v. *Industrial Tank Specialist Limited* felt able to say of the *Junior Books* decision:

> 'Faced with these difficulties it is, I think safest for this court to treat *Junior Books* as a case in which, on its particular facts, there was considered to be such a very close relationship between the parties that the defenders could, if the facts as pleaded were approved, be held liable to the pursuers.'

By 1983 architects, engineers, contractors and sub-contractors were at risk as to claims in negligence from a fairly wide range of potential plaintiffs. These included not only the people with whom they were in contract such as the developer, but also subsequent owners and occupiers, including tenants and sub-tenants. The criticism of this state of the law began to mount, particularly from the construction professions. **2.13**

The first steps on the road to retrenchment in the law of tort came with *Governors of the Peabody Donation Fund* v. *Sir Lindsay Parkinson & Co. Limited* in 1985. It was alleged in the case that the local authority owed a duty to Peabody in relation to drains which had to be reconstructed after they were found to be unsatisfactory. Lord Keith said: **2.14**

> 'The true question in each case is whether the particular defendant owed the particular plaintiff a duty of care having the scope which is contended for, and whether he was in breach of that duty with consequent loss to the plaintiffs. A relationship of proximity in Lord Atkin's sense must exist before any duty of care can arise, but the scope of the duty must depend on all the circumstances of the case.'

In relation to Lord Wilberforce's two stage test in *Anns* (see 2.11), Lord Keith also said: **2.15**

> 'There has been a tendency in some recent cases to treat these passages as being themselves of definitive character. This is a temptation which should be resisted.'

Of the same passage in *Anns*, Lord Brandon continued the retreat in *Leigh & Sillivan Limited* v. *Aliakmon Shipping Co. Limited* in 1986: **2.16**

'The first observation which I would make is that the passage does not provide, and cannot in my view have been intended by Lord Wilberforce to provide, a universally applicable test of the existence and scope of a duty of care in the law of negligence.'

2.17 The retreat from *Anns* continued in *Curran and Another* v. *Northern Ireland Co-ownership Housing Association Limited and Others* in 1987 where the court refused to hold that the Northern Ireland Housing Executive owed a duty of care to future owners of a house to see that an extension had been properly constructed.

2.18 The three cases, *Peabody*, *Aliakmon*, and *Curran*, have been referred to as the 'retreat from *Anns*' but in *Yuen Kun Yeu* v. *Attorney General of Hong Kong* in 1987, the two stage test of Lord Wilberforce in *Anns* was probably put to rest by Lord Keith. Further, in *Simaan General Contracting Co.* v. *Pilkington Glass* in 1988, the Court of Appeal held that a supplier of glass units for a new building who had no contractual relationship with the main contractor, and had not assumed responsibility to that contractor, was not liable in tort for foreseeable economic loss caused by defects in the units where there was no physical damage to the units, and the contractor had no proprietary or possessory interest in the property.

2.19 It follows from all this that the extensive duties in tort that had been developed in the 1960s, 70s and early 80s were in some disarray by 1988. The whole basis of the decision in *Anns* had received widespread criticism and it was inevitable that sooner or later a challenge was mounted in the House of Lords to their previous decision in *Anns*. The first opportunity was in *D & F Estates Ltd and Others* v. *Church Commissioners for England and Others* in 1988.

D & F ESTATES LTD AND OTHERS v. CHURCH COMMISSIONERS FOR ENGLAND AND OTHERS

2.20 The case concerned defective plastering carried out by sub-contractors to a main contractor. The non-occupying lease-holder, which was a company, claimed against the main contractor (with whom he did not at any time have a contract) in respect of costs of repair to plastering actually carried out, future repair costs and loss of rent. Lord Bridge delivered the main speech, the remainder of their Lordships agreeing with no substantial dissent.

2.21 The non-occupying lease-holder plaintiff had no option but to bring his case in tort against the main contractors for the simple reason that he had no contract with them. It was a difficult case to frame in the law of tort, if for no other reason because a contractor has no liability in law for the torts of his independent contractor, namely, the sub-contract plasterers. The plaintiffs therefore put their duty as a duty on the part of the main contractor to adequately supervise the work of the plastering sub-contractors. The judge at first instance found for the plaintiffs but the House of Lords overturned that decision and their reasons are of fundamental importance in the area of negligence liability in the construction industry. Consider these two passages from the speech of Lord Bridge in *D & F*:

'It is, however, of fundamental importance to observe that the duty of care laid down in *Donoghue* v. *Stevenson* was based on the existence of a danger of physical injury to persons or their property. That this is so is clear from the observations made by Lord Atkin at pages 581 to 582 with regard to the statements of law of Brett MR in *Heaven* v. *Pender* (1883). It has, further, until the present case, never been doubted so far as I know that the relevant property for the purpose of the wider principle on which the decision in *Donoghue* v. *Stevenson* was based was property other than the very property which gave rise to the danger of physical damage concerned'

and

'. . . there are two important considerations which ought to limit the scope of the duty of care which it is common ground was owed by the appellants to the respondents on the assumed facts of the present case. The first consideration is that, in *Donoghue* v. *Stevenson* itself and in all the numerous cases in which the principle of that decision has been applied to different but analogous factual situations, it has always been either stated expressly, or taken for granted, that an essential ingredient in the cause of action relied on was the existence of danger or the threat of danger or physical damage to persons or their property, excluding for this purpose the very piece of property from the defective condition of which such danger, or threat of danger, arises. To dispense with that essential ingredient in a cause of action of the kind concerned in the present case would, in my view, involve a radical departure from long established authority.'

The essence of what was being said was that the developments in the law of tort between 1932 and 1988 were tantamount to giving Donoghue, in   **2.22**

Donoghue v. *Stevenson*, not only damages for her personal injury in being made ill by the decomposed snail in the ginger beer bottle, but also requiring the manufacturer to pay for or provide a new bottle of ginger beer, the thing itself. On this basis, it was easy for the House of Lords in *D & F* to come to the view that the plaster, being the damaged thing itself, had not caused damage to persons or property (other than the *de minimis* cleaning of carpets involving an expenditure of about £50) and that the non-occupying lease-holder was not entitled to succeed against the contractor. However, in coming to that decision, the House of Lords had some difficulty in reconciling the *Anns* decision, although they did not overrule it. It follows from the *D & F* decision that, for example, tenants, purchasers and funds could not rely in future on the possibility of being able to obtain recompense in tort in respect of defects in design or construction of buildings; hence the immediate and urgent boost in the use of collateral warranties since that decision. The collateral warranty tries to fill the gap in the law of tort by creating a contractual relationship.

MURPHY v. BRENTWOOD DISTRICT COUNCIL

2.23   The decision in *Murphy* was delivered on 26 July 1990; it was widely known that in argument before the House of Lords, the local authority had asked the House to depart from its previous decision in *Anns* v. *Merton London Borough Council* — the House of Lords can overrule its previous decisions by reason of the Practice Statement (Judicial Precedent) [1966] 1 WLR 1234. Their Lordships, in some detailed judgments, reviewed the state of the law as it had developed since 1932 in relation to negligence, not only in England and Wales but also in the Commonwealth; they gave consideration to some American tort cases as well as looking at their own previous and recent decision in *D & F*. Lord Keith in *Murphy*, having expressly approved a passage in a case in the High Court of Australia, which declined to follows *Anns*, in *Council of the Shire of Sutherland* v. *Heyman*, said this:

'In my opinion, there can be no doubt that *Anns* has for long been widely regarded as an unsatisfactory decision. In relation to the scope of the duty owed by a local authority it proceeded upon what must, with due respect to its source, be regarded as a somewhat superficial examination of principle and there has been extreme difficulty, highlighted most recently by the speeches in *D & F Estates*, in ascertaining upon exactly what basis of principle it did proceed. I think it must now be recognised that it did not proceed on any basis of principle at all, but constituted a remarkable example of judicial legislation. It has engendered a vast spate of litigation, and each of the cases in the field which have reached this House has been distinguished. Others have been distinguished in the Court of Appeal. The result has been to keep the effect of the decision within reasonable bounds, but that has been achieved only by applying strictly the words of Lord Wilberforce and by refusing to accept the logical implications of the decision itself. These logical implications show that the case properly considered had potentiality for collision with long established principles regarding liability and the tort of negligence for economic loss. There can be no doubt that to depart from the decision would re-establish a degree of certainty in this field of law which it has done a remarkable amount to upset.'

Having then noted that the *Anns* decision had stood for some 13 years and that the House of Lords should be cautious in overruling their previous decisions, he said: **2.24**

'My Lords, I would hold that *Anns* was wrongly decided as regards the scope of any private law duty of care resting upon local authorities in relation to their function of taking steps to secure compliance with building by-laws or regulations and should be departed from. It follows that *Dutton* v. *Bognor Regis Urban District Council* should be overruled, as should all cases subsequent to *Anns* which were decided in reliance on it.'

The effect of this decision is therefore to substantially remove a cause of action in negligence which had been relied upon by tenants, subsequent owners and occupiers for a considerable period of time to enable them to recover damages in respect of negligent design and construction. Collateral warranties, to create a contractual relationship to fill this gap, are now regarded as being an essential matter as an adjunct to the development of commercial property. However, it may be that the *Murphy* case has put greater importance on to the decision of *Hedley Byrne & Co. Limited* v. *Heller & Partners Limited*; it may also be that *Junior Books Limited* v. *Veitchi Co. Limited*, which was regarded in *Murphy* as being an application of the *Hedley Byrne* principle, may also have been given something of a boost, notwithstanding the fact that in a great many recent cases, *Junior Books* has been heavily criticised. That **2.25** criticism was to be found, for example, in *D & F* where Lord Bridge said of *Junior Books*:

> 'The consensus of judicial opinion, with which I concur, seems to be that the decision of the majority is so far dependent upon the unique, albeit non-contractual, relationship between the pursuer and the defender in that case and the unique scope of the duty of care owed by the defender to the pursuer arising from that relationship that the decision cannot be regarded as laying down any principle of general application in the law of tort or delict.'

2.26 With respect to their Lordships, it is very hard for those involved in the workings of the construction industry to understand how the relationship between an employer and a sub-contractor is 'unique' or that the scope of the duty of care in that case could reasonably have been based on that unique relationship. There really can hardly be a more common relationship in the construction industry than that between an employer and a sub-contractor, nominated or otherwise. However, these issues, and in particular the concept of reliance, require further consideration as a result of the *Murphy* decision.

Hedley Byrne after Murphy

2.27 The wording adopted by their Lordships in *Hedley Byrne* was somewhat disparate. However in so far as it is possible to make one statement of principle from those judgments, that statement is set out at page 575 of the Law Report and is as follows:

> 'If, in the ordinary course of business or professional affairs, a person seeks information or advice from another, who is not under contractual or fiduciary obligation to give the information or advice, in circumstances in which a reasonable man so asked would know that he was being trusted, or that his skill or judgment was being relied on, and the person asked chooses to give the information or advice without clearly so qualifying his answer as to show that he does not accept responsibility, then the person replying accepts a legal duty to exercise such care as the circumstances require in making his reply; and for a failure to exercise that care an action for negligence will lie if damages result.'

2.28 In essence therefore *Hedley Byrne* was concerned with negligent mis-statements of facts or opinions, in such circumstances that it was reasonable to expect that the recipient of the information would rely on such information and the recipient did in fact rely upon the information. If these factors were present and the recipient suffered financial or economic loss then a claim in negligence could be brought against the person who gave the information. Such indeed were the facts of *Hedley Byrne* itself. A useful illustration of negligent mis-statement in the construction industry is in the *IBA* case the facts of which are set out in paragraphs 1.20 and 1.21 above. It will be recalled that BICC, who had no contractual relationship with IBA, were considered to be liable in respect of a negligent mis-statement arising from their letter to IBA dated 11 November 1964 when they negligently mis-stated to IBA that they were 'well satisfied that the structures will not oscillate dangerously . . .' IBA relied upon that assurance.

Clearly therefore the *Hedley Byrne* principle creates a particular species **2.29** of negligence based upon representation and reliance unconnected with physical damage and entitling a party to recover economic loss and is of particular relevance to advice or information given by professional people. Unlike *Anns* the species of negligence established by *Hedley Byrne* survives the *Murphy* decision. So far so good. The difficulties now start and they arise from the following extracts from the speeches of Lord Keith and Lord Bridge in *Junior Books Limited* v. *Veitchi Co. Limited*:

Per Lord Keith:

> 'It would seem that in a case such as [*Pirelli General Cable Works Ltd* v. *Oscar Faber & Partners*] where the tortious liability arose out of a contractual relationship with professional people, the duty extended to take reasonable care not to cause economic loss to the client by the advice given. The plaintiffs built the chimney as they did in reliance on that advice. The case would accordingly fall within the principle of *Hedley Byrne. I* regard *Junior Books Limited* v. *Veitchi Co. Limited as being an application of that principle.*'

Per Lord Bridge:

> 'There may, of course, be situations where, even in the absence of contract, there is a special relationship of proximity between builder and building owner which is sufficiently akin to contract to introduce the element of reliance so that the scope of the duty of care owed by the builder to the owner is wide enough to embrace purely economic loss. *The decision in Junior Books can, I believe, only be understood on this basis.*'

2.30   It is suggested that the problem arising from the speeches of Lord Keith and Lord Bridge is that in *Junior Books* the relationship between the employer and the sub-contractor is not a 'unique' relationship in the construction industry — indeed it is commonplace. Further there appears to be a complete absence of the representation and reliance that was central to the decision in *Hedley Byrne* and the court's approach in *IBA*. It is submitted that the references to 'reliance' in *Junior Books* were in the context of the test of proximity rather than a *Hedley Byrne* reliance. If, as their Lordships state, *Junior Books* is to be explained on the basis of *Hedley Byrne*, it seems to the authors that *Murphy* has succeeded in 'locking the front door but leaving open a rear window'. It would have been better, in the authors' view, if their Lordships had simply overruled *Junior Books* as they did *Anns* rather than try and justify the decision on the basis of *Hedley Byrne*. Only time will tell.

Chapter 3

Assignment and Novation

FUTURE PURCHASERS AND TENANTS

The Problems

The original developer and often the first purchaser or tenant of a **3.1** property will have an opportunity to enter into direct contractual arrangements to protect themselves against latent construction defects. It is unlikely that subsequent purchasers or tenants will have such an opportunity and, adopting the terminology of the rule of privity of contract, they will be strangers to the original contractual arrangements with no remedies in contract against the parties responsible for the design and construction of the building.

The courts have often said in such circumstances that the law of **3.2** contract provides for a chain of indemnity connecting the ultimate user with the original producer: for example D is the ultimate user or consumer who purchased from C the retailer, C having purchased from B the wholesaler and B having purchased from A the manufacturer. D can sue C for breach of contract but not B or A. However if C is sued by D, then C will have a right of indemnity against B, who in turn has a right of indemnity against A, creating 'the chain of indemnity' that links the manufacturer to the ultimate user. Unfortunately the strength of a chain of indemnity is only as great as its weakest link. If C the retailer becomes insolvent a critical link in the chain between D and A will have been broken. Further a purchaser of a freehold building is faced with the difficulty of the principle of *caveat emptor* and the tenant of a leasehold building with the difficulty of full repairing covenants in the lease. *Caveat emptor* and full repairing covenants are dealt with in Chapter 6.

The future purchaser or tenant must rely upon derivative contractual **3.3** rights. Such rights arise by assignment which is a unilateral act or by novation which is synallagmatic. (See paragraph 1.32 above.)

ASSIGNMENT

Assignment of Choses in Action

3.4 Choses or things in action are 'all personal rights of property which can only be claimed or enforced by action and not by taking physical possession': *Torkington* v. *Magee*. The term includes the benefits arising under a contract and, subject to certain qualifications dealt with in paragraph 3.31 below, rights of action arising by reason of a breach of contract. A chose in action can be a legal chose for example an interim payment due under a building contract or an equitable chose such as a legacy under a will or an interest in a partnership. Choses in action can be assigned or transferred unilaterally, for example A, the employer, enters into a construction contract with B, the contractor, requiring B to construct a building to a quality set out in the specification; A can, without the consent of or indeed knowledge of B, transfer the benefit of that contract to a third party C. A is known as the assignor, B the debtor and C the assignee. It is important to understand that the right to assign a chose in action is not derived from contract. It is a statutory right, alternatively a right arising from the rules of equity. Express conditions are commonly found in collateral warranties purporting to grant rights of assignment; these conditions are unnecessary and may have the effect of restricting the rights to assign.

Legal and Equitable Assignments

Common Law and Equity
3.5 There are four types of assignment namely:

(1) statutory or legal assignments of legal choses in action
(2) statutory or legal assignments of equitable choses in action
(3) equitable assignments of equitable choses in action
(4) equitable assignments of legal choses in action.

3.6 A brief knowledge of English legal history is helpful in understanding the dichotomy between legal and equitable assignments. Legal rights derive from the common law of England which was conceived and developed during the period between the Norman Conquest and the fourteenth century. The common law was administered by the king's justices on circuit through the three common law courts of King's Bench, Common Pleas and Exchequer. There were no courts of equity. However because of restrictions placed upon the continued development of the common law, not least being the baronial intimidation of the common law courts and their juries, plaintiffs in search of justice began to petition the king in council for a resolution of their disputes pursuant to the king's inherent judicial powers. Eventually this practice led to the petitions being referred to the king's chancellor who initially discharged this function in the name of the king but who subsequently established the Courts of Chancery as an independent tribunal from the king in council. The jurisdiction of the Courts of Chancery was based upon the canon law concept of 'conscience' and ultimately developed into the rules of equity. England therefore had two court systems namely the Common Law Courts and the Courts of Chancery each developing their own rules of law. This separation was abolished by statute in 1875 which replaced the old court structure with the present day structure of the Supreme Court of Judicature. Nevertheless the rules of equity remain distinct from the common law.

Legal Assignments of Choses in Action

The right to make a legal assignment is now governed by statute namely **3.7** section 136 of the Law of Property Act 1925. Sub-section 1 of section 136 provides as follows:

> 'Any absolute assignment by writing under the hand of the assignor (not purporting to be by way of charge only) of any debt or other legal thing in action, of which express notice in writing has been given to the debtor, trustee or other person from whom the assignor would have been entitled to claim such debt or thing in action, is effectual in law (subject to equities having priority over the right of the assignee) to pass and transfer from the date of such notice:
>
> (a) the legal right to such debt or thing in action;
> (b) all legal and other remedies for the same; and
> (c) the power to give a good discharge for the same without the concurrence of the assignor.'

It will be apparent from the wording of the sub-section that certain legal **3.8** formalities must be complied with if an assignment is to be an effective legal assignment. These formalities are as follows:

(1) an absolute assignment in writing signed by the assignor;
(2) a debt or other legal thing in action; and

(3) express notice in writing to the debtor.

3.9   An absolute assignment does not include the assignment of part of a debt
or thing in action, whether or not the part assigned is ascertained or
unascertained. In *Walter and Sullivan Limited* v. *J. Murphy & Sons
Limited* WS were plastering sub-contractors who commenced legal
proceedings against M for the sum of £1808 alleged to be due in respect
of a sub-contract for plastering works. After the commencement of the
proceedings WS, who were indebted to a third party H & Co., notified M
that M were 'to pay to H & Co. the sum of £1558 17s 8d from monies
owing by you to us . . . the receipt of H & Co. shall be good and sufficient
discharge to you in respect of payment made hereunder'. By a second
document H & Co. agreed with WS that in consideration of the
irrevocable authority given by them to M 'we will pay over to you any
monies which are paid to us by [the defendants] . . . after your debt to us
. . . has been fully repaid. . . .' The court held that the arrangement
between WS and H & Co. was an assignment of part of a debt and
therefore did not satisfy the requirements of sub-section 1 of section 136
of the Act.

3.10   An assignment which purports to be by way of charge only is not an
absolute assignment. This is a complex legal concept outside the ambit of
this book. Suffice it to say that the relevant test is to decide whether the
assignment merely gives a right to the assignee to payment out of a
particular fund by way of security rather than an unconditional transfer
of the fund to the assignee. In the *Walter and Sullivan* case the court also
held that, as well as being an assignment of part of a debt, the assignment
purported to be by way of charge. By way of contrast it was held in
Tancred v. *Delagoa Bay Company* that an assignment by way of mortgage
was absolute because there was a condition for re-assignment on payment
of the loan. It is the substance of the transaction and not the titles of
documents that determines the nature of the assignment.

3.11   An assignment which is qualified by conditions cannot be a legal
assignment. In *Williams, Williams* v. *Ball* the assignor purported to
transfer the benefit of a life insurance policy but made it conditional upon
the assignee surviving the assignor. This was held to be a conditional
assignment falling outside section 136 of the Act. The judicial reasoning
behind the requirement for an absolute assignment is that the debtor
should not be put in doubt or jeopardy by the arrangements between the
assignor and the assignee as to whom he is to discharge his obligations. In
the cases of *Walter and Sullivan* and *Williams* there were such doubts but
not in the case of *Tancred* where the re-assignment on repayment of the

loan would have to be notified to the debtor.

To create a legal assignment there must be a written document signed   **3.12**
by the assignor. Signature by an agent would not appear to be sufficient.
Any form of wording may be used provided there is a clear intention to
make an absolute assignment. The assignment may be a document
passing between the assignor and the assignee or a written demand from
the assignor to the debtor that the debtor pays or discharges his
obligations to the assignee. In the latter case in order to be an effective
assignment rather than merely an authority to pay a third party, there
must be evidence that the assignee consented to the arrangement between
the assignor and the debtor: *Curran* v. *Newpark Cinemas Limited.* Unlike
an assignment an authority to pay can be revoked prior to the actual
payment.

A debt or other legal thing in action includes both legal choses and   **3.13**
equitable choses. The purpose of section 136 of the Act which replaced
but substantially re-enacted section 25, sub-section 6 of the Judicature
Act 1873, was procedural and not intended to create new forms of choses
in action.

To create a valid legal assignment written notice of the assignment   **3.14**
must be given to the debtor. No particular form of wording is required,
indeed a document can constitute notice even though it was not intended
to be a notice. In *Van Lynn Developments Limited* v. *Pelias Construction
Co. Limited* P's bank overdraft was paid off by Van Lynn in consideration
of P assigning the debt to Van Lynn. The assignment was dated 26 June.
By a letter dated 27 June Van Lynn demanded payment from P. In their
letter Van Lynn stated, incorrectly, that notice of the assignment had
previously been given to P. The court held that a notice of assignment was
still good notice to the debtor even though it did not refer to the date of
the assignment. Further as regards Van Lynn's letter dated 27 June the
incorrect statement as to a notice could be ignored as 'an inaccurate
surplusage' and it was immaterial that the letter was not written with the
intention that it should perform the function of giving notice under the
Act. It is not necessary for the notice to the debtor to be given by the
assignor or the assignee; it may be given by a third party. In *Bateman* v.
Hunt a valid notice was given by the executor of a deceased sub-assignee.

Once there has been an assignment which complies with the formalities   **3.15**
of section 136 there is a transfer to the assignee of the legal right to the
chose in action and the assignee can give good discharge upon payment
or satisfaction by the debtor. It follows that the assignor has no right to
sue in respect of the chose in action unless of course there is a re-
assignment to the assignor. The same rules apply to intermediate

assignments thus creating a potential problem where a tenant assigns to a sub-tenant of part of the demised property.

3.16 An assignment within the statute does not require consideration thus voluntary assignments are enforceable between the assignor and the assignee and between the assignee and the debtor.

Equitable Assignments

3.17 A failure to comply with the formalities of section 136 of the Act is not necessarily fatal to the transaction; a defective legal assignment may operate as an equitable assignment: *William Brandts Sons & Co.* v. *Dunlop Rubber Co.* Indeed a defective legal assignment which takes effect as an equitable assignment may subsequently become a legal assignment if the defect is removed, for example where an equitable assignee of a defective legal assignment subsequently serves written notice on the debtor to perfect the legal assignment.

3.18 There may be an equitable assignment of an equitable chose or an equitable assignment of a legal chose. No consideration is required for the assignment of an equitable chose provided that the assignor has, at the material time, done all that he can do to perfect the gift: *Letts* v. *Inland Revenue Commissioners*. It is suggested that the better view is that the same rule applies to equitable assignments of legal choses although there are judicial *dicta* to the contrary.

3.19 An equitable assignment may be in writing or oral. Any words will suffice provided they are unambiguous. Referring to the form of an equitable assignment Lord Macnaghten in the *William Brandts* case stated:

> 'It may be addressed to the debtor. It may be couched in the language of commerce. It may be a courteous request. It may assume the form of mere permission. The language is immaterial if the meaning is plain. All that is necessary is that the debtor should be given to understand that the debt has been made over by the creditor to some third person.'

3.20 Lord Macnaghten's judgment in *William Brandts* referred to notice to the debtor. In law there may be a binding equitable assignment between assignor and assignee without notice to the debtor. However as a matter of practice notice to the debtor is very important for three reasons:

(1) In the absence of notice the debtor is entitled to discharge his obligations to the assignor and not to the assignee whereas if he has notice he does so at his own peril and he may well be required to

discharge the obligation a second time to the assignee with no entitlement to recovery from the assignor: *Walter and Sullivan Limited.*

(2) The giving of notice to the debtor has an effect upon prior equities. See paragraph 3.23 below.

(3) The date of notice establishes the order of priority as between successive assignees: *Dearle* v. *Hall.*

The notice may be written or oral and the wording of the notice informal although casual conversations may not be sufficient notice: *Re Croggon ex parte Carbis.* Indeed in the case of *Lloyd* v. *Banks* the court held that a newspaper article was sufficient notice to the debtor.

An equitable assignment may operate by way of a charge only or be of **3.21** part of the debt or chose: *Walter and Sullivan Limited.* Thus where a developer wishes to dispose of the completed building to more than one purchaser or tenant it is submitted that he will only be in a position to give each individual purchaser or tenant an equitable assignment of the benefits arising under the principal design and construction contracts. If a legal assignment is required then the draftsman of the principal contracts should take care to impose an obligation on the designers and contractors to provide a sufficient number of collateral warranties to satisfy the requirements of multi-occupation.

Procedural Differences between Legal and Equitable Assignment

Substantively legal and equitable assignments (provided notice has been **3.22** given to the debtor) are the same. There are however important procedural differences. As previously stated a legal assignment within the Act transfers a legal right in the chose to the assignee. Consequently the assignee sues the debtor in his own name. If there is an equitable assignment of an equitable chose in action the assignment being absolute, then again the assignee is entitled to sue in his own name. However if there is an equitable assignment of a legal chose in action or an equitable chose which is not absolute, for example a part of the debt, the assignor must be joined into the action either as plaintiff, if he co-operates, or as defendant if he does not. If the assignor is not joined as a party the assignee's action may well fail, although it is important to stress that these requirements are procedural and are not substantive; therefore the courts have a discretion to dispense with joinder of the assignor if they are satisfied that there is no prospect of a further claim by the assignor: *The Aiolos.* Also note that Order 15 Rule 6 of the Rules of the Supreme Court provides that an action shall not be defeated by reason of the non-joinder of a party although the court has power to direct that additional parties should be joined.

PRIOR EQUITIES

3.23   The effect of an assignment, whether it is a legal assignment or an equitable assignment, is to place the assignee in the shoes of the assignor in respect of the benefits (but not the burdens for which see paragraph 3.43 below) arising under the original transaction with the debtor. Consequently the assignee cannot by the assignment obtain a more advantageous position *vis à vis* the original debtor than that which was occupied by the assignor: *Business Computers Limited* v. *Anglo African Leasing Limited* where Templeman J stated that:

> 'A debt which accrues due before notice of an assignment is received, whether or not it is payable before that date, or a debt which arises out of the same contract as that which gives rise to the assigned debt, or is closely connected with that contract, may be set off against the assignee.'

3.24   It is important to note that if the set-off arises independently from the original contract between the assignor and the debtor then it cannot be set off against the assignee if the liability, as distinct from the actual payment, accrued after the date of receipt of a notice of assignment. The giving of notice of assignment is however irrelevant to claims by way of set-off or counterclaim that arise from the original contract or a contract which is closely connected to the original contract. For example A is the developer, B the architect appointed by A, C the first purchaser of the development from A and D the second purchaser from C. The contract between A and B provides for design works to be carried out by B and payment therefor to be made by A. B also enters into a collateral warranty undertaking to C that he will carry out his design works with reasonable skill and care. C assigns the benefit of the collateral warranty to D. A has not paid all of B's professional fees. In the event that B is in breach of his collateral warranty if D brings proceedings against B then B will be able to set off the amount of the unpaid fees against D's claims regardless of whether the entitlement to the fees arose after the date of D's notice of assignment to B. This is because the collateral warranty and the original contract between A and B are closely connected contracts. In the above example the same right of set-off arises as between B and C if C were the ultimate purchaser who took an assignment of A's benefits under the original contract with B. In this latter example the rights of set-off and counterclaim would arise from the same contract.

A counterclaim for unliquidated damages i.e. damages which have not yet been quantified may be set off by the debtor against any claims brought by the assignee: *Phoenix Assurance Co. Limited* v. *Earls Court Limited.* **3.25**

The debtor's right to counterclaim against the assignee is limited to defending the claims brought by the assignee, the counterclaim being set off in extinction or diminution of the assignee's claims. It does not entitle the debtor to bring positive counterclaims against the assignee i.e. for sums in excess of the assignee's claims. This is because, as stated above, the assignee only takes the benefit and not the burden of the original contract. **3.26**

Intermediate Assignees

It would appear that where there have been successive assignments the debtor is not entitled to set off against claims brought by the ultimate assignee counterclaims which the debtor has against intermediate assignees: *The Raven.* **3.27**

RESTRICTIONS ON ASSIGNMENT

The right to assign may be governed by specific statutory provisions; for example section 53(1)(c) of the Law of Property Act 1925 provides that equitable assignments of interest in a trust must be in writing. Other examples are bills of lading, copyright, patent rights and life insurance policies. If there is a statutory code this will override the general provisions of section 136 of the Act and also the rules of equity. **3.28**

Future Choses in Action

Future choses in action cannot be assigned either by a legal assignment or by an equitable assignment. The distinction between an existing chose and a future chose is not as obvious as it sounds and can often lead to complex forensic analysis. For example a future right to payment under an existing contractual right will be an existing chose not a future chose. In comparison the benefits which are likely to flow from a contract not yet entered into will be a future chose. Further difficulties arise in respect of **3.29** accrued contractual obligations which may be defeated by a subsequent breach of the contract by the party otherwise entitled to the accrued benefit. In *Hughes* v. *Pump House Hotel Company Limited* it was held that a contractor's right to be paid under a building contract was an existing chose even though the right of payment might be defeated by the contractor's subsequent failure to perform his contractual obligations.

3.30   The vital difference between an existing chose and a future chose is that the latter can only be enforced if there is an agreement supported by adequate consideration. In *Re McArdle (deceased), McArdle* v. *McArdle* M and his wife lived in a dwelling house forming part of the estate of M's father in which M and his brothers and sisters were beneficially interested expectant on the death of the tenant for life. In 1943 and 1944 Mr and Mrs M carried out certain improvements and decorations in and on the house, the cost of which, amounting to £488, was borne by Mrs M. In April 1945 M and his brothers and sister signed a document addressed to Mrs M which provided:

'In consideration of your carrying out certain alterations and improvements to [the dwelling house] at present occupied by you, we the beneficiary under the Will of [the father] hereby agree that the executor . . . shall repay to you from the said estate when so distributed the sum of £488 in settlement of the amount spent on such improvements.'

In 1948 the tenant for life died; Mrs M claimed payment of the sum of £488. The court held that the consideration for the execution of the document in April 1945 was past consideration and the document could not operate as an equitable assignment for valuable consideration. Further whilst an equitable assignment could be valid without consideration the document did not constitute such an assignment because, as it contemplated future action by Mrs M to the satisfaction of the signatories, it did not render her title complete and so was not complete and perfect. In other words the court found that the document purported to be an assignment of a future chose in action therefore it would not be binding unless it constituted an agreement supported by valuable consideration (see paragraphs 1.42 to 1.44).

Bare Rights of Action

3.31   Legal or equitable assignments of bare rights of action, that is to say litigation, are void and unenforceable. This is because the courts consider such transactions to be trading in litigation or to use the lawyer's terminology 'offending against the rules of maintenance and champerty'. In *Prosser* v. *Edmonds* Lord Abinger CB stated:

'It is a rule not of our law alone, but of that of all countries, that the mere right of purchase shall not give a man a right to legal remedies. The contrary doctrine is nowhere tolerated and is against good policy. All our cases of maintenance and champerty are founded on the principle that no encouragement should be given to litigation by the introduction of parties to enforce those rights which others are not disposed to enforce.'

Maintenance is an agreement to fund litigation and champerty (a species **3.32** of maintenance) is an agreement to divide the proceeds of the litigation. Prior to the Criminal Law Act 1967 maintenance and champerty were both criminal acts and torts. Section 14(2) of the Act abolished the criminal offence and also the tort of maintenance and champerty. However it preserved the rule of law that agreements that tended to either maintenance or champerty would be void and unenforceable.

Clearly the benefits arising under a contract do not offend against the **3.33** above rule: see paragraph 3.4. What is the position however if prior to a purported assignment of the benefits of a contract, those benefits have crystallised into rights of action for damages for breach of contract, for example latent construction defects? The courts had to consider a similar situation in *Dawson* v. *Great Northern and City Railway Co*. The Railway Company constructed under statutory powers a tunnel under or near certain houses in which D was interested and in which she carried on business and she claimed to be entitled to compensation on the ground that her interest in the houses had been injuriously affected by structural damage to the houses and by damage to trade stock. The structural damage had occurred prior to the acquisition of her freehold and leasehold interests although both transactions purported to assign rights to recover compensation from the Railway Company in respect of the structural damage to her properties, these rights being statutory rights under the Land Clauses Consolidation Act 1845. The court held that whilst an assignment of a mere right of litigation is bad, an assignment of property is valid even though that property may be incapable of being recovered without litigation. Sterling LJ stated:

'Even if the assignment be regarded apart from the conveyance of the lands and buildings . . . it appears to us that it is good; but we think that great weight must be given to the circumstances that this assignment is incidental and subsidiary to that conveyance and is part of a *bona fide* transaction the object of which was to transfer to the plaintiff the property of [the vendor] with all the incidents which attach to it in his hands. Such a transaction appears to be very far removed from being a transfer of a mere right of litigation.'

3.34 A similar issue arose in *Ellis* v. *Torrington*. The facts were somewhat complex but illuminating. The property in question was subject to three leases namely a head lease expiring on 25 December 1917, an underlease expiring on 18 December 1917 and a sub-underlease expiring on 15 December 1917. All these three leases contained onerous covenants to repair. The defendant T was the tenant pursuant to the sub-underlease and E the plaintiff was a sub-tenant of T, although E's covenants to repair were far less onerous than those imposed upon T. On 18 December 1917 E purchased the freehold interest of the premises which was subsequently conveyed to him together with the benefits of the covenants to repair in the head lease. At the expiration of all these leases the premises were substantially out of repair. T threatened E with proceedings on the basis of E's covenant to repair in his sub-tenancy from T. Faced with this threat E, who could not pursue T under the head lease, obtained an assignment of the benefits of the covenants in T's sub-underlease and then commenced proceedings against T for breach of covenant as assignee of the sub-under lessor. The court held that the assignment was free from objection on the ground of maintenance or champerty, the right of action on the covenants being so connected with the enjoyment of property as to be more than a bare right to litigate. Banks LJ stated:

'The respondent is seeking to enforce a right incidental to property, a right to a sum of money which theoretically is part of the property he has bought.'

3.35 More often than not with construction projects the assignments of benefits arising under collateral warranties or collateral contracts, even though they have crystallised into rights of action, will be incidental to property and therefore falling within the principles set out in *Dawson* and *Ellis*. In any event in recent years the courts have become more relaxed about the issue of maintenance and champerty. In 1968 Lord Denning stated:

'Much maintenance is considered justifiable today which would in 1914 have been considered obnoxious': *Hill* v. *Archbold*.

The leading case is the House of Lords' decision in *Trendtex Trading Corporation and Another* v. *Crédit Suisse*. Again the facts of this case were somewhat complex involving proceedings in England in respect of an agreement made in Switzerland containing a purported assignment by a Swiss corporation to a Swiss bank of its rights of action against the Central Bank of Nigeria in respect of a dishonoured letter of credit for the sum of US$14,000,000. The bank had assigned to an undisclosed third party all the rights of action against the Central Bank of Nigeria (CBN) for the sum of US$1,100,000. Less than five weeks from the date of the assignment the third party had settled the claim against CBN upon payment by the bank of US$8,000,000. One of the issues in the case was the validity of the intermediate assignment to the Swiss bank and the subsequent assignment to the third party. The court held that in determining the validity of an assignment of a cause of action it was the totality of the transaction that was to be looked at and *if the assignment was of a property right or interest and the cause of action was ancillary to that right or interest, or if the assignee had a genuine commercial interest in taking the assignment and enforcing it for his own benefit, the assignment would not be struck down as an assignment of a bare cause of action or as savouring of maintenance.*

Accordingly if no parties other than Trendtex and Crédit Suisse had been involved the intermediate assignment would have been valid even though it involved an assignment of Trendtex's residual interest in the CBN litigation as Crédit Suisse had a genuine and substantial interest in the success of that litigation. However the introduction of the third party rendered the agreement void under English law because the agreement clearly showed on its face that its purpose was to enable the cause of action against CBN to be sold to an anonymous third party with the likelihood of that third party, which had no genuine commercial interest in the claim, making a profit out of the assignment. Thus *Trendtex* introduced a further qualification to the rule against the validity of assignments of bare rights of action, namely where there is a genuine commercial interest in taking the assignment which of course does not necessarily have to be an interest incidental to the use of property.

Personal Contracts

If the contract is a personal contract then the benefits of that contract 3.36 cannot be assigned either by a legal assignment or by an equitable assignment. The test, which is an objective test, is whether 'it can make no difference to the person on whom the obligation lies to which of two persons he is to discharge it': *Tolhurst* v. *Association Portland Cement Manufacturers Limited.* It will be noted that the test is not concerned with the personal skill of the debtor. For example if an architect gives a collateral warranty relating to his design work, whilst the design work will involve personal skill on his part and will not be assignable by the architect to a third party, the benefits of the undertakings arising under the collateral warranty given to, say, a tenant will be assignable to a future tenant because, applying the objective test, it can make no difference to the architect whether his obligations lie to the first tenant or to the future tenant. It could however make a difference if a subjective test were the proper test, for example the future tenant could be more litigious than the existing tenant and more likely to bring proceedings against the architect in the event of breach.

3.37   There is a presumption in favour of commercial contracts that the benefits arising under such contracts are assignable. However the presumption is rebuttable. In *Kemp* v. *Baerselman* B contracted with K, a cake manufacturer, to supply him with all the eggs of a specified quality 'that he shall require for manufacturing purposes for one year'. K undertook not to purchase eggs from any other merchant during the year so long as B was ready to supply them. During the relevant year K transferred his business to a company whereupon B claimed to be discharged from his contract and refused to supply any more eggs to K or to the new company. The court held that B's contract was with K personally and that the benefit of the contract was not assignable. The court considered that the personal characteristics of the contract were twofold namely that B's obligations were defined by reference to K's manufacturing purposes and further that K had undertaken not to purchase eggs from any other merchant.

In *Tolhurst* the owner of land had contracted with a company to supply them for 50 years with at least 750 tonnes of chalk per week and so much more as they might require for their manufacture of cement. The original company was a small business which went into voluntary liquidation and transferred all its assets, including an assignment of the benefit of the supply contract, to Associated Portland Cement. Associated Portland Cement was a much larger concern than the original company and carried on business at various places. The court held that the assignment was effective. In *Kemp* Lord Alverston CJ distinguished *Tolhurst* on the basis that in the latter case the House of Lords had 'treated the contract as a supply to a given cement making place and not as a personal contract'.

Conditions of Contract

Often the contract between the assignor and the original debtor will **3.38**
include an express term purporting to prohibit or restrict the right of
assignment. See for example clause 19.1 of the JCT Form of Contract
1980 Edition and clause 12 of the RIBA Form of Collateral Warranty
(now treated by the RIBA as being superseded by the BPF Warranty).
Are such conditions effective? Some earlier cases appear to suggest that
the right of assignment cannot be prohibited or restricted by the
contractual arrangements between the assignor and the original debtor.
In *Tom Shaw & Co.* v. *Moss Empires (Limited) and Bastow* B was a comedy
artist who appeared on stage for a season at the Moss Empires Theatre.
His contract with Moss Empires provided by clause 13 that he should not
assign his salary which should be paid direct to him and to no other
person except in the case of his death. His booking at the Moss Empires
Theatre was obtained through his agents Tom Shaw & Co. with whom B
agreed that:

'I hereby agree to pay you or your assigns 10% commission on [my]
salary and on all monies which should accrue under the said
engagement or a prolongation of the same... and hereby authorise
Moss Empires to deduct and pay the said commission from my salary
in any manner which you may deem expedient.'

Notice of this letter was given by Tom Shaw to Moss Empires. The court
held that despite the restriction in B's contract with Moss Empires, B's
letter to Tom Shaw & Co. constituted a valid equitable assignment.
Darling J stated:

'The strongest ground for the defence was in the contract . . . clause 13.
But though Moss Empires might bring an action for breach of that
contract, if they could show any damages . . . it could no more operate
to invalidate the assignment then it could to interfere with the laws of
gravitation.'

The strange thing about the *Tom Shaw* case was that the judge awarded **3.39**
damages and costs against B and not Moss Empires! Also Moss Empires
admitted that they were liable to pay the plaintiff. In *Spellman* v. *Spellman*
there was conflicting *obiter dicta*. Danckwerts LJ considered that 'the fact
that there is a prohibition in the document creating the chose in action
against assignment is not necessarily fatal to such claim'. However in the

same case Willmer LJ considered that any prohibitions should be binding as 'it would be quite impossible for this court to make an order to the contrary because such an order would in effect require the husband to break his contract with the hire purchase corporation'.

3.40   It is suggested that the better view is the view expressed in *Helstan Securities Limited* v. *Hertfordshire County Council* where the court held that if the parties to a contract, the subject matter of which was a chose in action, agreed that the chose in action was not to be assigned any purported assignment was invalid. In this case Hertfordshire County Council entered into a civil engineering roadworks contract with a contractor. The contract was the ICE Conditions of Contract, Fourth Edition. Condition 3 provided: 'The Contractor shall not assign the contract or any part thereof or any benefit or interest therein or thereunder without the written consent of the employer.' The contractor got into financial difficulties and without obtaining the Council's consent assigned to Helstan the amount of £46,437 allegedly to be owing by the Council to the contractor. The Council refused to discharge the amount to Helstan who brought proceedings against the Council for payment. Croom-Johnson J stated:

> 'If the reported cases are not a sure guide, one is thrown back in this case on the agreement. There are certain kinds of choses in action which, for one reason or another, are not assignable and there is no reason why the parties to an agreement may not contract to give its subject matter the quality of unassignability.'

3.41   The Judge distinguished the *Tom Shaw* case on its own peculiar facts and also on the basis that, if that case was good authority for the proposition that a contractual prohibition against an assignment was not effective, the principle was limited to the relationship between assignor and assignee and not between assignee and original debtor.

3.42   It is to be noted however that the judgment in *Helstan Securities* placed an emphasis upon contracts. It is submitted that attempts to prohibit or restrict assignability must satisfy the requirements of a contract and clauses which give an unconditional right to a first assignment but then endeavour to restrict future assignments may well fail as they could be considered to be in breach of the rule of privity of contract. There is also the difficulty that the assignee steps into the shoes of the assignor who has of course an unconditional right to assign, *ad infinitem*.

NOVATION

As stated above it is only the benefits of a contract which can be   **3.43**
transferred by way of an assignment. If the parties wish to transfer both
the benefit and the burdens then this must be done by a novation
agreement.

A novation occurs when there is a rescission of one contract and the   **3.44**
substitution of a fresh contract in which the original contractual
obligations are carried out by different parties. For example A is a
developer who enters into a contract with B the contractor. Before
practical completion of the construction works and by way of a separate
transaction A agrees a sale of the completed project to C. To give C the
benefit of the contractual specification for the building works, A, B and
C may enter into a novation agreement which substitutes C for A in the
original contract with B. If it is a true novation agreement then its effect
will be to release A from all liability in respect of the original contract and
C may be sued by B for any breaches of contract by A pre-existing the
date of the novation agreement. As a matter of practice however express
terms are often introduced into novation agreements with a view to
restricting the retrospective characteristics of novation. Such qualified
agreements are really variation agreements rather than novation agreements, and such agreements must be supported by adequate consideration. With a novation agreement the consideration is deemed to be the
mutual discharge of the old contract: *Scarf* v. *Jardine.*

A novation agreement is not possible without consent. It is essential   **3.45**
therefore that the principal contracts between developer and consultants
and between developer and contractors contain express terms obliging
the contractor and the consultant to enter into the novation agreement.
To avoid the risk of merely having an agreement to agree, which is
unenforceable, it is suggested that a specimen form of novation
agreement be appended to the principal contractual documentation.

Chapter 4

Reasonable Skill and Care and Fitness for Purpose

4.1   One of the common problems associated with collateral warranties is disconformity between the performance obligations set out in the principal contract and the performance obligations in the warranty itself. The former usually provides for a performance obligation of reasonable skill and care and the latter often attempts to impose a performance obligation of fitness for purpose.

REASONABLE SKILL AND CARE

4.2   A contractual obligation to carry out works or services with reasonable skill and care creates a performance obligation which is analogous to the standard of care in negligence. The court considered the standard of care in negligence in *Blyth* v. *Birmingham Waterworks Co.* and stated:

> 'Negligence is the omission to do something which a reasonable man, guided upon those considerations which ordinarily regulate the conduct of human affairs, would do, or doing something which a prudent and reasonable man would not do.'

4.3   *Blyth* established the appropriate tests for the behaviour of the general public and not for the behaviour of members of a more limited group who have or hold themselves out as having specialist skills such as architects or engineers. In *Bolam* v. *Friern Hospital Management Committee*, approved in *Whitehouse* v. *Jordan*, the court refined the test established in *Blyth* in order to accommodate specialist skills. The court applied the following test:

> 'Where you get a situation which involves the use of some specialist skill or competence, then the test of whether there has been negligence

or not is not the test of the man on the top of a Clapham omnibus because he has not got this special skill. A man may not possess the highest expert skill at the risk of being found negligent. It is well established law that it is sufficient to be exercising the ordinary skill of an ordinary competent man exercising that particular art.'

The facts of *Bolam* were concerned with a medical negligence case, **4.4** nevertheless the test set out in the judgment is equally applicable to other professional men and those exercising specialist skills: *Greaves & Co. (Contractors) Limited* v. *Baynham Meikle and Partners.*

An error of judgment or the selection of the wrong method, where **4.5** there is a genuine difference of specialists' opinion, will not necessarily amount to negligence. In *Robinson* v. *The Post Office* R, a doctor in general practice, injected a patient with an anti-tetanus serum without first administering a test dose. At the relevant time, medical opinion was moving against the use of anti-tetanus serums generally. The Court of Appeal held that since R's failure to give a test dose was contrary to accepted procedure, he had been negligent, but that no damage had been caused as the result of the test would have been negative, and that, since at the time there was still a responsible body of medical opinion who favoured anti-tetanus serum, he had not been negligent in using it. Similarily, in the case of *Perry* v. *Tendring District Council and Others* which dealt with, *inter alia*, the failure of a consultant engineer to design foundations that would be unaffected by long term soil heave, the court considered that the standard of care depended upon 'what was to be expected of the competent engineer at the material time' (i.e. the time he designed the foundations). This is the 'state of the art' defence. In *Perry* there was conflicting expert witness evidence. One expert engineer personally knew of heave but was only able to refer to one text book intended for engineers which dealt with it. Another expert had not read that text book and thought that engineers generally would not have known of heave although that particular expert had expressed a contrary view some 12 years after the material time. Two other engineers stated categorically that they had never heard of heave. Judge Newey stated:

'On the totality of the expert evidence I must, however reluctantly, conclude that at the material time a competent engineer would not have known of long term heave.'

A more draconian attitude was adopted by the House of Lords in the *IBA* **4.6** case, the facts of which are set out in paragraph 1.20 above. The defendant's submissions that the design and building of the cylindrical mast was work which was 'both at and beyond the frontier of professional knowledge at that time' was not disputed by their Lordships; nevertheless they held that the designer was negligent. As regards the state of the art submission, Viscount Dilhorne stated:

> 'No doubt all this was true, and bearing in mind the consequences that might ensue if such a mast collapsed — fortunately no-one was killed or injured at Emley Moor, though part of the mast fell across a road and it might have fallen on a farmhouse — it was in my opinion incumbent on [the designer] to exercise a very high degree of care.'

4.7   There are conflicting judicial decisions on the issue of whether a professional man who specialises within his profession has a higher duty than the non-specialist. In *Wimpey Construction UK* v. *Poole* a consultant held himself out as having especially high skills and was retained on that basis. The court rejected the argument that the test in such circumstances should be that of a man exercising or professing to have especially high professional skills. However in *Ashcroft* v. *Mersey Regional Health Authority* the court found that the more skilled a person is, the more is the care which is to be expected of him, but the test should be applied without gloss either way. The authors presume that the reference to 'gloss' was to indicate a pragmatic, rather than strict, application of the test.

4.8   Competency will invariably be a matter of expert evidence and opinion. However, in the last resort, the courts consider that they have a discretion to reject expert evidence as to what is an acceptable practice within a profession. In *Sidaway* v. *Governors of Bethlem Royal Hospital* Lord Templeman stated:

> 'Where the practice of the medical profession is divided or does not include express mention, it will be for the court to determine whether the harm suffered is an example of a general danger inherent in the nature of the operation, and if so whether the explanation afforded to the patient was sufficient to alert the patient to the general dangers of which the harm suffered is an example.'

4.9   *Sidaway* was concerned with a surgeon's duty to warn a patient of a potential risk, and on that basis can be distinguished from a designer of a building project. However, it is suggested that the court would have a similar discretion in construction cases. Indeed, the House of Lords adopted a similar position in the *IBA* case.

The express conditions of the principal contract or the collateral **4.10**
warranty may determine the standard of performance. What, however, is
the position if the contract is silent on this particular point? In the *Greaves*
case, there was a suggestion that a designer's obligation might extend
beyond reasonable skill and care to fitness for purpose. The facts of the
Greaves case concerned G, a building contractor who undertook to
design and construct on a package deal basis, a new factory, warehouse
and offices for Alexander Duckham Limited. The warehouse was to be
used for the storage of barrels of oil. G contracted with B, structural
engineers, to design the structure of the warehouse. G informed B that the
floors of the warehouse had to take the weight of forklift trucks carrying
barrels of oil. After completion and occupation, cracks began to appear
in the floors of the warehouse. It was established that the failure of the
floors was due to vibrations caused by the use of the forklift trucks. The
issues before the court turned upon whether B were in breach of their
obligation to carry out their design works with reasonable skill and care
or whether B were in breach of an implied term of the contract between
G and B that B's design should be fit for its purpose, namely the
movement of loaded forklift trucks. Implied terms are discussed in
paragraphs 1.65 to 1.67. It will be appreciated that a term which is implied
as a matter of fact does not have the consequences of a term that is implied
as a matter of law, insofar as the former only relates to the particular
bargain struck between the parties to the contract, whereas the latter
applies to all bargains unless excluded by the express terms of the contract
or the circumstances surrounding the making of the contract.

In the *Greaves* case, the judgment at first instance appeared to suggest
that a fitness for purpose obligation was to be implied as a matter of law.
On first reading, this is also the impression given by the judgment of
Denning MR in the Court of Appeal. However, on its facts, the *Greaves*
case does not create a universal principle of fitness for purpose on the part
of designers, in that the court found that, whilst there was a contractual
term that the designers should design a warehouse that was fit for its
purpose, this term was implied as a matter of fact and not law. The court
also found that B's design was negligent, that is to say in breach of the
obligation to carry out their services with reasonable skill and care.

Any doubts that lingered from the *Greaves* case were disposed of by the **4.11**
Court of Appeal in *George Hawkins* v. *Chrysler (UK) Limited and Burn
Associates*. B were engineers who contracted with C to prepare the design
and specification for a shower room at C's factory which included a new
floor and wall coverings. G, the plaintiff, was an employee of C and he
slipped on a puddle of water in the shower room after having used the

shower. G sued C and C in turn brought proceedings against B. C settled G's claim but continued the third party proceedings against B. The main issues in the case were:

(1) was there an implied term of the contract that B would use reasonable skill and care in selecting the material to be used for the floor of the shower room? and
(2) was there an implied warranty or term that the material used for the floor would be fit for use in a wet shower room?

The judge at first instance found against C in respect of the first issue, but on the second issue the judge found that B were in breach of an implied warranty that they would provide 'as safe a floor as was practicable in the expertise of the profession to provide a safe floor for these men in these conditions'. On appeal, the Court of Appeal held, *inter alia*, that although a party contracting for both the design and supply of a product will usually be under an implied contractual duty to ensure that it is reasonably fit for the purpose for which it is intended, where the contracting party is a professional man providing advice or designs alone (i.e. without supplying any product), no warranty will normally be implied beyond a term that reasonable skill and care will be taken in giving the advice or preparing the design. There was nothing in the present case to require the implication of any term other than a duty to take reasonable care and skill in preparing the design.

4.12   If the party being called upon to enter into a collateral warranty is an architect or engineer with a design function, and his principal contract expressly provides for a performance obligation of reasonable skill and care or is silent on this matter, for that party to enter into a collateral warranty with a fitness for purpose obligation then he will be increasing his potential liabilities. The important distinction between reasonable skill and care and fitness for purpose is that fitness for purpose is an absolute obligation, and provided the obligation is clearly established or defined by the contract document, the party in breach will not be able to plead as a defence that he has discharged his services with reasonable skill and care. The facts in *Samuels* v. *Davis* provide a useful illustration of the dichotomy between reasonable skill and care and fitness for purpose. In *Samuels*, the Court of Appeal held that where a dentist undertakes for reward to make a denture for a patient, there is an implied term of the contract that the denture will be reasonably fit for its intended purpose. Du Parcq LJ stated:

'If someone goes to a professional man . . . and says: "Will you make me something which will fit a particular part of my body?" and the professional gentleman says "Yes", without qualification, he is then warranting that when he has made the article, it will fit the part of the body in question. . . . If a dentist takes out a tooth or a surgeon removes an appendix, he is bound to take reasonable care and to show such skill as may be expected from a qualified practitioner. The case is entirely different where a chattel is ultimately to be delivered.'

The words of Du Parcq LJ were cited with approval by Lord Scarman in the *IBA* case.

It is important to note that if an architect or engineer extends his **4.13** potential liability by entering into a collateral warranty providing for a fitness for purpose obligation, there may well be serious repercussions in respect of his professional indemnity policy (see paragraphs 7.30 to 7.32 below).

FITNESS FOR PURPOSE

The first section of this chapter has dealt with the legal obligations of a **4.14** professional man. What is the position of the contractor or sub-contractor, in particular the design and build contractor?

Clearly, as stated above, if the contract expressly deals with the **4.15** standard of the contractor's performance then, in the absence of ambiguity, the express terms will determine the extent of the contractor's or sub-contractor's legal obligation. However, if the contract is silent on these matters it has long been held that a contractor or sub-contractor who agrees to carry out construction works impliedly warrants (that is to say there is term implied by law) that he will carry out his works with reasonable skill and care (often referred to as the obligation to carry out the works in a good and workmanlike manner). The standard of performance is the same as reasonable skill and care in negligence.

The contractor or sub-contractor also warrants that the materials he supplies for the purposes of works will be of a merchantable quality, that is to say good of their kind. This warranty is an absolute warranty and extends to latent defects and it will not help the contractor or sub-contractor to show that he has exercised reasonable skill and care in the selection of those materials: *Young & Marten Limited* v. *McManus Childs Limited*. M were developers of a residential housing estate and Y were a firm of roofing sub-contractors. Y provided an estimate for the supply and laying of certain roof tiles subsequent to which M specified that Y should use a particular roof tile called 'Somerset 13'. These tiles were supplied by only one manufacturer J. Beale & Co. The tiles supplied by Beale appeared to be sound. However, 12 months after completion of the roofs a large number of tiles began to disintegrate — a consequence of a latent defect. M were sued by the purchasers of the houses and M sought indemnity against Y. At first instance, the court rejected M's submission that there was an implied term that the Somerset 13 tiles should be reasonably fit for their purpose and should be of merchantable quality. On appeal the Court of Appeal held:

(1) Unless the circumstances of a particular case suffice to exclude, then there will be implied into a contract for the supply of work and materials a term that the materials used will be of merchantable quality and a further term that the materials used will be reasonably fit for the purpose for which they are used.

(2) In this particular case the circumstances sufficed to exclude the term that the tiles would be reasonably fit for the purpose for which they were required.

(3) In this particular case the circumstances were not sufficient to exclude the term that the tiles were merchantable. The fact that these tiles were obtainable from only one manufacturer was not a circumstance which excluded the implication but, *per* Lord Reid, if the tiles had been made by only one manufacturer who was willing to sell only on terms which excluded or limited the ordinary liability (under statute) and if that fact was known to the employer and to the contractor when they made the contract, then it would be unreasonable to place upon the contractor a liability for latent defects.

(4) Y supplied and fixed tiles which were latently defective and thereby breached the implied term (of merchantable quality).

4.16 In *Gloucestershire County Council* v. *Richardson*, the House of Lords found that the particular circumstances of the case excluded both the implied warranty of suitability and the implied warranty of merchantable quality. In that case R entered into a contract with G for the construction of an extension to a technical college. The contract was in the RIBA Form 1939 Edition, 1957 Revision. The bills of quantities provided for a prime cost sum for concrete columns to be supplied by a nominated supplier. R contracted to erect the columns. Clause 22 of the conditions of contract dealing with nominated suppliers (unlike clause 21 which dealt with nominated sub-contractors) did not entitle R to make reasonable objection to a proposed supplier, nor to object on the ground that the supplier would not indemnify him in respect of his main contractor's obligation. G's architect instructed R to accept a quotation given by C W & Co for the supply of the concrete columns. CW's standard conditions of trade restricted their liability in respect of good supply by them. The columns supplied by CW had latent defects because of faulty manufacture and after erection cracks appeared in them; the columns were unsuitable for use as structural members of the extension. The House of Lords considered that the circumstances set out above indicated an intention on the part of G and R to exclude from the main contract any implied terms that the concrete columns should be of good quality and fit for their required purpose.

Supply of Goods and Services Act 1982

The common law rules have now in part been replaced (as to goods) and **4.17** in part supplemented (as to services) by a statutory code, the Supply of Goods and Services Act 1982 which governs all contracts made after 4 January 1983. Section 4 of the Act sets out the code for quality and fitness. Section 4(2) provides that goods shall be of a merchantable quality save as to defects which have been notified or which should have been apparent on examination before the contract was made. Sections 4(4) and 4(5) provide that where the purpose for which the goods are being acquired is made known by the party contracting to buy the same, whether expressly or by implication, there is an implied condition that the goods should be reasonably fit for their purpose, whether or not that is a purpose for which such goods are commonly supplied. Section 4(6) creates a proviso whereby the fitness for purpose condition does not apply if the other party does not rely, or it was unreasonable for him to rely, on the skill or judgment of the party providing the goods.

Section 4(9) provides that goods shall be of merchantable quality 'if **4.18** they are fit for the purposes or purposes for which goods of that kind are commonly supplied as it is reasonable to expect having regard to any description applied to them, the price (if relevant) and all other relevant circumstances'. For example, goods which are seconds do not have to be of the same quality as goods which are described as first class.

Section 1(3) of the Act provides that a contract is contract for the **4.19** transfer of goods even though services are to be provided under the same contract. It follows that section 4 applies to the materials part of a contract for the supply of work and materials. A building contract is a contract for the supply of work and materials. A design contract is a contract for services.

4.20   Section 13 of the Act deals with a contract for services (i.e. a design contract or the work content of a building contract) and provides that there shall be an implied term that the contractor will carry out the services with reasonable care and skill. Unlike the statutory warranties as to quality and fitness of goods section 13 does not exclude the common law rules insofar as those rules impose a stricter duty than the Act. It follows that the decisions in *Young & Marten* and *Gloucestershire* are still of relevance.

Design and Build Contractors

4.21   What it is the standard of performance of the contractor who undertakes a design obligation? Is it the same as the professional man, that is to say reasonable skill and care, or is it the higher duty of fitness for purpose? This issue was considered in the *Greaves* case. Lord Denning stated:

> 'Now, as between the building owners and the contractors, it is plain that the owners made known to the contractors the purpose for which the building was required, so as to show that they relied on the contractors' skill and judgment. It was, therefore, the duty of the contractors to see that the finished work was reasonably fit for the purpose for which they knew it was required. It was not merely an obligation to use reasonable care. The contractors were obliged to ensure that the finished work was reasonably fit for the purpose.'

4.22   In *Viking Grain Storage Limited* v. *T. H. White Installations Limited*, W were package deal contractors for the design and construction of a grain drying and storage installation. The installation was not fit for its purpose and V contended that there were implied terms of the contract that W would use materials of good quality and reasonably fit for their purpose and that the completed works would be reasonably fit for their purpose, namely that of a grain drying and storage installation. The court held that there were no terms of the contract or any other relevant circumstances which were inconsistent with the implied terms of quality and fitness for purpose, and further that there was no reason to differentiate between W's obligation in relation to the quality of materials and their obligation as to design. V had relied upon W in all aspects, including design and on the skill and judgment of W, and in the circumstances the terms contended for should be implied.

By way of contrast the Supreme Court of Ireland, in the case of *Norta Wallpapers (Ireland)* v. *Sisk & Sons (Dublin)*, held that where a roof structure, which had been supplied and erected by a specialist sub-contractor, subsequently leaked and was unsuitable for its purpose, the fact that the main contractor was given no choice but to use the specialist sub-contractor, his design and price constituted circumstances which meant that there was no reliance by the employer on the main contractor and, accordingly, there was no fitness for purpose obligation on the main contractor in respect of the specialist sub-contractor's failure.

In the recent (unreported) case of *Trolex Products Limited* v. *Merrol Fire Protection Engineers Limited* there was an interesting issue as to whether a design obligation was created by the bringing together of what otherwise would have been standard components. T were the sub-contractors for the supply of an electronic control system which was incorporated into M's own works, comprising the installation of a fire protection system in the Ras Abufontas power and water station in Qatar. In response to the submission by T that there was minimal design obligation in the sub-contract Potter J stated: **4.23**

'I should perhaps add that at one stage I had evidence from T which minimised the work of design carried out, suggesting that it was no more than, in effect, a matching of pieces of standard equipment to make up a package to do the job; or as Mr B put it "logic design work created from standard equipment". Even if that was so in fact, I am satisfied from the answers of Mr B that there was a conscious realisation that design work was involved and that T were consulted as experts in their field. Further, it is clear that substantial time was spent on this work. Again, whether or not that was so, it is not suggested that anything was said by T to delimit or belittle the design work involved and the construction of the written contract is clear in my view, namely as one for work of design as well as the supply of goods.'

DWELLINGS

Where a contractor is involved in the construction of a residential dwelling there is an implied term, implied by law, that the contractor will carry out his work in a good and workmanlike manner, that he will supply good and proper materials and that the dwelling will be reasonably fit for human habitation: *Hancock* v. *B. W. Brazier (Anerley) Limited*. This common law obligation has now been supplemented by a **4.24** statutory code set out in the Defective Premises Act 1972 which came into force on 1 January 1974. Section 1(1) of the Act provides:

> 'A person taking on work for or in connection with the provision of a dwelling (whether the dwelling is provided by the erection or by the conversion or enlargement of the building) owes a duty,
>
> (a) if the dwelling is provided to the order of any person, to that person; and
> (b) without prejudice to paragraph (a) above to every person who acquires an interest (whether legal or equitable) in the dwelling;
>
> to see that the work which he takes on is done in a workmanlike or, as the case may be, professional manner with proper materials and so that as regards that work the dwelling will be fit for habitation when completed.'

4.25 Unlike the Supply of Goods and Services Act discussed above, the Defective Premises Act does not operate via the parties' contract but in fact, creates a statutory duty. Accordingly, it has a much wider effect and is akin to tortious liability which is not dependent upon a contract. It follows that future owners are entitled to sue the builder if he is in breach of the statutory duty. The Act is, however, restricted to the provision of dwellings and does not apply to commercial developments although the term 'dwellings' includes dwellings that are created by conversion or enlargement. There is also a statutory exception; section 2(1) provides that where the construction of the dwellings is subject to 'an approved scheme' the Act does not apply. The National House-Building Council (NHBC) operates a warranty scheme for dwellings. The NHBC's schemes dated 1973, 1975, 1977 and 1979 are approved schemes under the Act. See the House Building Standards (Approved Scheme) Order 1979 Statutory Instrument 1979/381. As far as the authors are aware, the most recent NHBC scheme, introduced in March 1988 is not yet an approved scheme under the Act.

4.26 The Act prohibits any attempt to exclude its operation by section 6(3) which makes void any term in a contract which purports to exclude or restrict the operation of the Act cf. the Supply of Goods and Services Act which can be excluded by express agreement, course of dealings or custom and usage [Sections 11 and 16].

Chapter 5

Damages and Limitation of Action

DAMAGES

Nature of Damages

Damages for breach of contract are intended to be compensatory, that is **5.1** to say so far as money can, they are intended to place the innocent party in the same position that such party would have been had the other party performed his contractual promises: *British Westinghouse Electric Company Limited* v. *Underground Electric Railways*. In contrast an award of damages in respect of tortious liability is intended to place the innocent party back in the position he was before the breach of the duty. Awards of damages for breach of contract may therefore be greater than awards of damages in tort. In *Muirhead* v. *Industrial Tank Specialities Limited* M entered into a contract with a third party for the supply of pumps to be used at M's lobster farm. The electric motors for the pumps were supplied by ITS. The motors failed causing the pumps to fail which resulted in the death of M's lobsters. There was no contract between M and ITS and M brought proceedings in negligence. M claimed three heads of damage: namely the value of the dead lobsters, the loss of profits on those lobsters and the loss of profits that would have been earned by M's business had the pumps functioned properly. It was held that M was entitled to recover the first two heads of damage but not the third head of damage. The court considered that the third head of damage might have been recoverable had there been a contract between M and ITS.

Damages may be general damages which can be assessed in an **5.2** approximate figure or special damages which represent past pecuniary loss and have to be calculated and pleaded with sufficient particularity so as not to take the defendant by surprise at trial. Further damages may include future damages, that is to say those which it is anticipated are likely to flow from the breach of contract. Future damages are known as prospective loss and must be claimed at the same time as past or present loss in respect of the same breach of contract: *Conqueror* v. *Boot*. This rule does not apply to breaches of different promises within the same contract or breaches of recurring obligations; such breaches are considered to give rise to separate causes of action. Damages for breach of contract (which term includes liquidated and ascertained damages) must be distinguished from claims arising under the terms of the contract, for example a claim for direct loss and expense under clause 26 of JCT 80. Such claims are not subject to the general principles on damages set out below; they are however subject to the rules of construction of a contract dealt with in paragraphs 1.50 to 1.59.

5.3   The parties to a contract may agree on the amount of damages that are to be awarded in the event of a breach of contract. Such damages are known as liquidated and ascertained damages and are common place in construction contracts, usually related to the completion obligations of the contractor. Provided the agreed sum is a genuine pre-estimate of loss and not a penalty, it will be enforced by the courts: *Dunlop Limited* v. *New Garage Co. Limited*. Damages may also include consequential losses. Consequential losses, often referred to as 'economic loss', are losses, such as loss of profit, which are not directly related to physical damage: *Muirhead*. See also paragraphs 8.74 and 8.75.

Causation

5.4   There must be a causal link between the breach of contract and the damage suffered by the innocent party. In *Quinn* v. *Burch Brothers (Builders) Limited* Q was a sub-contractor to B in respect of certain building works. There was an implied term of the sub-contract that B would supply, within a reasonable time, any plant or equipment reasonably necessary for carrying out the sub-contract works. B was in breach of that term in failing to supply a step ladder requested by Q. To prevent any delay Q used a trestle which he knew to be unsuitable unless it was footed by another workman. The trestle was not footed and it moved causing injury to Q. It was held that B's breach of contract provided the occasion for Q to injure himself but was not the cause of his injury which was caused by his own voluntary act in using the trestle; accordingly B was not liable to pay damages for Q's injury for his damage was not a natural and probable consequence of the breach of contract even if, as was in fact doubtful, it was a foreseeable consequence of that breach.

5.5   The causal link may be broken by the intervening acts of the plaintiff, as in *Quinn's* case, or by the intervening acts of a third party. It must be noted however that the courts are wary of laying down any general principles and many cases turn upon their own particular facts. For example in the case of *Weld Blundell* v. *Stephens* S negligently and in breach of contract permitted a libellous letter written by W to fall into the hands of a third party who communicated the contents of the letter to the person who had been libelled. That person sued W who in turn sued S. The court held that the acts of the third party broke the causal link between the S's initial breach and the W's damage.

Where there are intervening *events* as distinct from intervening acts of **5.6** the plaintiff or a third party, such events will not break the causal link between breach and damage if they were foreseeable. In *Monarch Steamship Co. Limited* v. *Karlshamms Oljefabriker (A/B)* the defendant shipowners chartered a ship to the plaintiffs for the carriage of a cargo from Manchuria to Sweden on terms that the voyage would be completed by July 1939. In breach of the charter party the ship was unseaworthy resulting in a delay to the voyage and in September 1939, on the outbreak of the Second World War, the British Admiralty diverted the ship to Glasgow resulting in transshipment of the cargo to Sweden on neutral vessels involving additional cost. The court held that in July 1939 the parties to the charter party should have had in mind the possibility of the outbreak of war and have foreseen the events which subsequently happened. Accordingly the defendant was entitled to recover damages for breach of contract.

Measure of Damages

The term 'measure of damages' can be used either in a wider sense to **5.7** include both categories of damage and the calculation or quantification of damage, or in a more restricted sense, namely the principles of law which define the categories or heads of damage which will be recoverable when there has been a breach of contract, commonly referred to by lawyers as the 'rules of remoteness of damage'. It is the latter restricted meaning which is adopted for the purposes of this sub-section.

The basic rule as to measure of damages is often referred to as the rule **5.8** in *Hadley* v. *Baxendale*. This was the name of a case heard in 1854 involving a claim for breach of contract by a mill owner against a carrier and arising from the carrier's failure to deliver a crankshaft within the time specified by the contract of carriage. Unbeknown to the carrier the crankshaft was critical to the whole of the output of the mill. The plaintiffs brought a claim against the defendants claiming a loss of profit for the whole of their production between the dates when the crankshaft should have been delivered and the date when it was actually delivered. In rejecting their claim for loss of profits Alderson B stated:

'Where two parties have made a contract which one of them has broken, the damages which the other party ought to receive in respect of such breach of contract should be such as may fairly and reasonably be considered either as arising naturally i.e. according to the usual course of things, from such breach of contract itself, or such as may reasonably be supposed to have been in the contemplation of both parties, at the time they made the contract, as the probable result of the breach of it.'

5.9 The judgment in *Hadley* v. *Baxendale* was explained and indeed developed in two leading cases in the twentieth century namely *Victoria Laundry (Windsor) Ltd* v. *Newman Industries Ltd* and *Koufos* v. *Czarnikow Ltd (The Heron II)*. In the *Victoria Laundry* case the rule was explained with reference to three main propositions namely:

'(1) In cases of breach of contract the aggrieved party is only entitled to recover such part of the loss actually resulting as was at the time of the contract reasonably foreseeable as liable to result from the breach.

(2) What was at that time reasonably foreseeable depends on the knowledge then possessed by the parties or, at all events, by the party who later commits the breach.

(3) For this purpose knowledge "possessed" is of two kinds; one imputed, the other actual. Everyone, as a reasonable person, is taken to know the "ordinary course of things" and consequently what loss is liable to result from a breach of contract in the ordinary course. . . . But to this knowledge, which a contract breaker is assumed to possess whether he actually possesses it or not, there may have to be added in a particular case knowledge which he actually possesses, of special circumstances outside the "ordinary course of things" of such a kind that breach in those special circumstances would be liable to cause more loss. . . .'

5.10 The wording of the judgment in *Hadley* v. *Baxendale* caused much lively debate on the issue of whether there were two branches to the rule, the first branch being damages arising naturally and the second branch being damages in the reasonable contemplation of the parties. This issue was not purely academic for it will be appreciated that if the first branch of the rule was unqualified by the parties' reasonable contemplation, or to put it another way by the particular bargain struck between the parties, then the first branch of the rule would tend towards the reasonable foreseeability test applicable to the measure of damages in tort, that is to say damage which should have been foreseen by a reasonable man as being something of which there was a real risk: *The Wagon Mound (No. 2)*. However the judgments in *Victoria Laundry* and *Koufos* support the view that there is really only one rule in *Hadley* v. *Baxendale* and that damages which may reasonably be supposed to have been contemplated by the contracting parties are damages which arise naturally from a breach of contract. What was in the reasonable contemplation of the parties is to be decided on both an objective basis and a subjective basis. The objective test turns upon the contemplation of a reasonable person, that is to say, it is imputed knowledge, whereas the subjective test turns upon the actual knowledge of the parties or the particular party who is in breach of contract.

The application of the rule in *Hadley* v. *Baxendale* can be usefully **5.11** illustrated by reference to the facts of the *Victoria Laundry* case and the *Koufos* case.

Victoria Laundry

V entered into a contract to purchase from N, an engineering firm, a **5.12** boiler which was installed on N's premises. V, who were a firm of launderers and dyers, required the boiler to extend their business to increase their general turnover and also they had in mind the prospect of certain profitable dying contracts being obtained from the Government. V and N agreed that the boiler should be delivered to V's premises on 5 June 1946. The boiler was damaged whilst it was being dismantled on N's premises and delivery to V was delayed until 8 November 1946. It was found as a matter of evidence that N were aware not only of the nature of V's business but also that V intended to put the boiler into use as quickly as possible. V sued N for breach of contract and their claim for damages included their loss of business profits.

At the trial at first instance the judge disallowed the claim for loss of profits on the ground that it was based upon special knowledge which had not been drawn to the attention of N. V appealed and the Court of Appeal, reversing the trial judge's decision, held that N, an engineering company with knowledge of the nature of the plaintiff's business, having promised delivery by a particular date of a large and expensive piece of plant, could not reasonably contend that they could not foresee that loss of business profit would be liable to result to the purchaser from a long delay in delivery; and that although N had no knowledge of the dying contracts which V had in prospect, it did not follow that V was precluded from recovering some general and perhaps conjectural sum for loss of business in respect of the contracts to be reasonably expected. The Court of Appeal was applying the objective test of reasonable contemplation as regards the general business profits of the launderers. However the specific profits which would have been earned from the prospective dying contracts with the Government were not claimable as N did not have actual knowledge of those contracts.

Koufos

5.13. C was the owner of the *SS Heron II*. C entered into a charter party with K for the consignment of sugar from Constanza to Basrah. At the time of making the contract the ship was docked in Piraeus and the shipowners anticipated that it would be ready to load in Constanza by about 25 to 27 October 1960 after which date it would proceed at all convenient speed to Basrah. The ship arrived at Constanza on 27 October, was loaded with sugar and departed from that port on 1 November. A reasonably accurate prediction of the length of the voyage between Constanza and Basrah was 20 days. C knew that K were sugar merchants and that there was a sugar market in Basrah but C had no actual knowledge that K intended to sell the sugar promptly after its arrival. In breach of the charter party contract the *SS Heron II* deviated from the voyage by calling at Berbera, Bahrain and Abadan, delaying the voyage by some 10 days, the ship arriving at Basrah on 2 December and not 22 November. The prices on the sugar market at Basrah tended to fall in October and November to a low point in December. Between 22 and 28 November the price of sugar in the Basrah market was £32 10s per tonne and the price between 2 and 4 December was £31 2s 9d per tonne.

K brought a claim against C for breach of contract claiming damages based upon a loss of profit by reason of the price differential. The House of Lords held that, since prices in a commodity market were liable to fluctuate, shipowners should reasonably contemplate that it was not unlikely that, if their ships delayed their voyage, the value of marketable goods on board their ships would decline and that therefore where there was wrongful delay in the delivery of marketable goods under a contract of carriage by sea the measure of damages was the difference between the price of the goods at their destination when they should have been delivered and the price when they were in fact delivered. Again the court was applying the objective test and did not feel it necessary to consider the actual knowledge of the contract breaker.

EXPECTATION INTEREST AND RELIANCE EXPENDITURE

Expectation Interest

The rules as to measure of damages give rise to two broad categories of **5.14** damage namely expectation interest and reliance expenditure. Expectation interest is the primary category of damage and represents the difference between the value to the promisee of a promise which has been performed satisfactorily and the value to the promisee of a promise which has been performed defectively or incompletely. The loss of profits on the sale of the sugar in the *Koufos* case is an example of expectation interest. Similarly where there are building defects the expectation interest will be the diminution in value, that is to say, the difference between the market value of the property without defects and the market value of the property with the defects.

Reliance Expenditure

Save for claims against building surveyors where the measure of damage **5.15** is diminution in value (see *Perry* v. *Sidney Phillips & Son*), the measure of damage in respect of building defects will often be based upon reliance expenditure and not expectation interest. A reliance expenditure award of damages is payment of compensation for wasted expenditure incurred by the promisee in reliance on the promisor's promise to perform. There are four broad categories of reliance expenditure:

(1) expenditure incurred by the promisee in order to perform his part of the contract for example the cost of labour or materials
(2) expenditure incurred by the promisee prior to entering into the contract but which will be wasted if there is not a satisfactory performance of the contract, for example the taking on of additional laundry staff in the *Victoria Laundry* case.
(3) expenditure incurred prior to the breach of contract not related to the promisor's performance but which will be wasted if the promisee fails to fulfil his contractual obligations, for example in the *British Westinghouse* case the plaintiffs incurred the costs of extra coal used by defective generators
(4) expenditure incurred after the breach of contract, for example the costs of repairs paid to a third party.

It is the fourth category of expenditure which is most commonly awarded **5.16** in the construction cases as the measure of damage for defective buildings and is often referred to by lawyers as 'substituted performance'. In *East Ham Corporation* v. *Bernard Sunley & Sons Limited* B constructed a school for E under the then current RIBA Form of Contract. Several years after completion of the work stone panels fixed to the exterior walls fell off owing to defective fixing. The court held that the proper measure of damages was the cost of replacing the stone panels.

5.17   As a general rule the person entitled to damages for breach of contract may elect to recover expectation interest or reliance expenditure. However where the damages arise from a breach of a construction contract, because of defective building works or incomplete building works, the courts appear to require the plaintiff to demonstrate an intention to carry out the works or intention to continue to occupy the relevant building or premises. The proper measure of damage is more than an academic debate. It can have a profound effect on the damages recovered particularly when the costs of carrying out the remedial works or completing outstanding works can be substantially greater than the diminution in the market value of the property.

Such were the circumstances in *Radford* v. *DeFroberville and Lange*. R was the owner of a large house in Holland Park, London which was let into flats. The house had a large garden and R decided to sell for building purposes part of the garden which fronted on the highway. R obtained planning permission to build a new house on the proposed plot and he agreed to sell the undeveloped site to D. In the document of transfer D agreed to build a new house in accordance with the planning permission and also to construct a wall in accordance with the detailed specification along the boundary of R's retained land and the new plot, the wall to be situated on the new plot. These obligations had to be carried out within a certain time which was subsequently extended. However D eventually sold the plot to L and the proposed house and the dividing wall remained unbuilt. R sued D for breach of contract because of D's failure to construct the dividing wall. D admitted liability but denied that R had suffered any loss on the basis that the proper measure of damage was diminution in value and there was no diminution as the absence of a physical barrier on the boundary of the property was unlikely to bring about any significant diminution in letting value. The cost of erecting the wall was £3400. The court held that as R intended to make good D's breach by building a wall himself on his own property the proper measure of damages was the cost of carrying out the work on his own land.

5.18   In the *Radford* case the defendant had failed to carry out or complete the works. In *Harbutts Plasticine Limited* v. *Wayne Tank & Pump Co. Limited* (overruled by *Photo Production* v. *Securicor Transport* on a different point) the defendants W were responsible for defective work. W entered into a contract with H to design and install equipment for storing and dispensing stearine in a molton state (at temperatures of between 120°F and 160°F) at H's factory which was an old building. W specified Durapipe, a form of plastic pipe, which was to be heated by electrical tapes wound round the pipe controlled by a thermostat. In fact Durapipe was wholly unsuitable for this purpose because it was liable to distort at temperatures of about 187°F and had a low thermal conductivity. The installation was completed on 5 February 1963 and both parties intended to test it the next day. As it was very cold, to ensure that the stearine would be molten for the test, an employee of W switched on the heating tapes on the night of 5 February and the installation was left unattended during that night. In the early hours of 6 February there was a fire which destroyed the factory. H rebuilt the factory but because of planning restrictions they had to replace their old five storey mill with a new two storey factory. The costs of building the new factory was £67,973 compared with the diminution in value of the old mill before and after the fire of £42,538. The court held that the proper measure of damage was the cost of reinstating the factory, not the difference in its value before and after the fire. Widgery LJ stated:

'The distinction between those cases in which the measure of damage is the cost of repair of the damaged article, and those in which it is the diminution in value of the article, is not clearly defined. In my opinion each case depends on its own facts. . . . If the article damaged is a motor car of popular make, the plaintiff cannot charge the defendant with the cost of repair when it is cheaper to buy a similar car on the market. On the other hand, if no substitute for the damaged article is available and no reasonable alternative can be provided, the plaintiff should be entitled to the costs of repair. It was clear in the present case that it was reasonable for the plaintiffs to rebuild their factory, because there was no other way in which they could carry on their business and retain their labour force.'

In *Tito & Others* v. *Waddell & Others No. 2* (*The Ocean Island* case) the **5.19** court held that a plaintiff can establish that his loss consisted of or included the costs of doing the work if he could show that he had done the work, or intended to do it, even though there was no certainty that he would.

Claims for both Expectation Interest and Reliance Expenditure

5.20   To what extent can a party claim both expectation interest and wasted reliance expenditure? The Court of Appeal has held that a party cannot claim both heads of damage; he must seek either expectation interest or wasted reliance expenditure: *Cullinane* v. *British 'Rema' Manufacturing Co. Limited.* It may be however that *Cullinane* can be distinguished on the basis that in that case the plaintiff claimed an expectation interest based upon his gross profits and not net profits; if the claim is limited to net profits there is no duplication of damages. Similarly where there is a claim for damages for building defects it would appear arguable that a claim for the costs of repairs together with a claim that the property had suffered a diminution in its market value because of the fact that it is a repaired structure, are not overlapping claims for damages and both heads of damage should be recoverable. Indeed both costs of repairs and diminution in market value were recovered in the case of *Thomas & Others* v. *T. A. Phillips (Builders) Limited and Taff Ely Borough Council.* This case involved negligence and not breach of contract. However the decision on measure of damages should also be relevant to a breach of contract case.

MITIGATION AND ASSESSMENT

5.21   Lawyers often refer to the plaintiff's duty to mitigate his loss. To talk about duty is probably adopting too high a standard of conduct; it is more helpful to consider mitigation in terms of reasonableness.

> 'A plaintiff was under no duty to mitigate his loss, despite the habitual use by lawyers of the phrase "duty to mitigate". He was completely free to act as he judged to be in his best interest. On the other hand, a defendant was not liable for all the loss suffered by the plaintiff in consequence of his so acting. A defendant was only liable for such part of the plaintiff's loss as was properly to be regarded as caused by the defendant's breach of duty': *Sotiros Shipping Inc. and Another* v. *Sameiet Solholt.*

Essentially therefore a plaintiff will not be allowed to recover damage which could have been avoided had the plaintiff acted reasonably. The burden of proof rests upon the defendant to show that the plaintiff has behaved unreasonably. The level of behaviour is one to be decided on the facts of each particular case although as a general rule the courts tend to favour the plaintiff and are often unimpressed with defendant's attempts to demonstrate that, with the benefit of hindsight, the plaintiff's behaviour was unreasonable. For example the courts do not expect a plaintiff to do anything other than that which is in the ordinary course of a business: *Dunkirk Colliery Co.* v. *Lever*.

If a plaintiff's reasonable attempts to mitigate the loss fail and result in **5.22** additional loss or damage such losses or damage may be recoverable from the defendant: *Banco De Portugal* v. *Waterlow & Sons Limited*. However if the plaintiff takes greater steps than he need have done and these result in a reduction of the loss and damage, then the defendant is entitled to the benefit of that reduction.

Mitigation has been described as the 'mirror image' of the rules of **5.23** remoteness discussed above and also the rules of assessment discussed below in paragraph 5.26. That is to say, the courts often disregard strict application of the rules and are more concerned to answer what has been described as the real question, namely what is the loss to the plaintiff. 'In the end the question seems to me to come down to a very short point. The cost is a loss if it is shown to be a loss': *per* Megarry VC: *The Ocean Island Case*.

Betterment

Betterment is a topic of particular relevance to defective building works **5.24** and involves consideration of both measure of damage and mitigation. In the *Harbutts Plasticine* case betterment was considered under the heading of measure of damage. The court held that the plaintiffs were not required to give credit for betterment merely because they had replaced the old building with a new one of modern design. Widgery LJ stated:

'The plaintiffs rebuilt their factory to a substantially different design, and if this had involved expenditure beyond the costs of replacing the old, the difference might not have been recoverable, but there is no suggestion of this here. Nor do I accept that the plaintiffs must give credit under the heading of "betterment" for the fact that their new factory is modern in design and material. To do so would be the equivalent of forcing the plaintiffs to invest their money in the modernising of their plant which might be highly inconvenient for them.'

In *Governers of the Hospital for Sick Children and Another* v. *McLaughlin* **5.25** *and Harvey plc and Others* the issue of unnecessary expenditure was approached from the point of view of mitigation. The defendants endeavoured to argue that despite the plaintiffs' reliance on expert opinion the plaintiffs had not acted reasonably in selecting their repair scheme. The defendants' argument was rejected by the court.

Assessment

5.26   The appropriate date for assessment of damages is often considered adjunctively with mitigation. The general rule is that damages shall be assessed at the date of breach of contract: *Miliangos* v. *George Frank (Textiles) Limited.* However as a matter of practice the courts' approach is invariably based upon a finding as to the reasonableness of the plaintiff's conduct particularly where the claim is for wasted reliance expenditure, for example the cost of building repairs. In the *East Ham Corporation* case the court held that the cost of repairing the stone panels should be assessed at the time when the defects were discovered and put right. In that case the breaches of contract occurred in May 1954 and the remedials were carried out in 1960. The case of *Bradford* adopted the test that the costs of repairs should be assessed at the date when in all the circumstances it was reasonable for the plaintiffs to commence the repairs. In *Dodd Properties Limited* v. *Canterbury City Council* (a case involving the tort of nuisance but of guidance to the issue of the date of assessment of the costs of building repair works) the court considered it reasonable for a plaintiff to delay the carrying out of necessary repair costs until such time as a judgment had been awarded against the defendant. These cases would appear to support the argument that impecuniosity or lack of credit may be one of the factors which can be taken into account in deciding whether or not the plaintiff has behaved reasonably in postponing the carrying out of repairs. It is a question of whether in all the circumstances it was 'a matter of commercial good sense' to delay the carrying out of the remedial works. This is to be contrasted with the situation where impecuniosity or lack of funds is the sole reason for not carrying out the remedial works; in such circumstances the increased costs of the remedial works may well be irrecoverable: *Liesbosch, Dredger* v. *Edison SS.*

Difficulties of Ascertainment

5.27   Provided the head of damage satisfies the test of remoteness the courts will award damages for breach of contract even though the precise quantification of the loss is not possible. The courts will make approximate assessments of damages. What is the position however if that assessment is dependent upon a contingency? For example an undertaking given by a developer to a consultant in a collateral warranty agreement that the developer will obtain collateral warranties in like terms from all the other consultants involved in the development. If the developer is in breach of his undertaking, will damages be recoverable by the consultant? This situation has often been described as the loss of a chance or opportunity.

The question of whether damages could be claimed in respect of a loss **5.28** of a chance was considered in *Chaplin* v. *Hicks*. H, a well known actor and theatrical manager, published a letter in a London daily newspaper stating that, with a view to dealing once and for all with the numerous applications he received from young ladies desirous of obtaining engagements as actresses, he was willing that the readers of the newspaper should by their vote select twelve ladies to whom he would then give engagements. The ladies were to apply by way of a photograph endorsing their name and address on the rear face of the photographs. H received a tumultuous response, some 6,000 ladies, whereby he decided to pre-select some 50 ladies on a regional basis from which the readers of the newspaper would then select the final 12. C, the plaintiff, was one of the ladies who went forward to the pre-selection stage. However by reason of some confusion about her address she was not notified in time to attend the pre-selection interview and her photograph was not put forward to the final selection by the readers of the newspaper. C sued H for breach of contract and it was argued on behalf of H that, even if there had been a breach, C's claim for damages was speculative as there was no certainty that she would be one of the twelve ladies selected by the readers of the newspaper. This argument was rejected by the court and it was held that where by a contract a person has a right to belong to a limited class of competitors for a prize, a breach of that contract, by reason of which that person is prevented from continuing a member of the class and is thereby deprived of all chance of obtaining the prize, is a breach in respect of which the person may be entitled to recover substantial and not merely nominal damages. *The existence of a contingency which is dependent on the volition of a third person does not necessarily render the damages for a breach of contract incapable of assessment.*

In *Cook* v. *Swinfen* (not followed in *Midland Bank* v. *Hett Stubbs & Kemp* but not on this point) S was a solicitor who was acting for C, the wife, in divorce proceedings. The husband had brought divorce proceedings against C and S was negligent in not defending the husband's petition nor cross petitioning for divorce on the basis of the husband's adultery. The husband's petition was unopposed and C brought proceedings against S claiming damages for the loss of a chance to obtain a divorce against her husband and maintenance for herself and her child. The court held that C was entitled to recover such damages. The court recognised that, in so far as it was anticipating the outcome of hypothetical proceedings brought by C against her former husband, to that extent the court's assessment was speculative; nevertheless the court considered that it was in a position to make an assessment based upon the balance of probability. It follows, in the authors' submission, that a court would be faced with similar speculation if in the example set out in paragraph 5.27 above the developer failed to obtain collateral warranties from other consultants. The court would have to consider whether on the balance of probabilities there was a likelihood of those other consultants being responsible for defective design which would have given rise to a right of contribution between all the consultants. It is presumed that in the light of the decisions in *Chaplin* and *Cook* the courts would be prepared to award damages in such circumstances.

ASSIGNMENT

5.29 The topic of damages merits individual consideration in respect of assignment of choses in action. There would appear to be two important questions in relation to collateral warranties the answers to which have been troubling the construction industry, namely:

   (1) Does the assignee have any right to recover damages against the original debtor if the assignor has not incurred any expenditure on repair or rebuilding costs or has sold his property for its full market value?
   (2) Does the assignee have any right to recover against the original debtor greater damages than would have been recoverable by the assignor?

Question 1

5.30 In the *Ocean Islands* case Megarry VC stated:

'If the plaintiff has suffered little or no monetary loss in the reduction of value of his land, and he has no intention of applying any damages towards carrying out the work contracted for, or its equivalent, I cannot see why he should recover the costs of doing work which will never be done. It would be a mere pretence to say that this cost was a loss and should be recoverable as damages.'

In *Perry* v. *Tendring* (on its facts based upon a claim in negligence but of some relevance on the issue of damage) the court expressly doubted whether an assignor who had sold for full market value had suffered any loss. Judge Newey stated *obiter*: **5.31**

'I am also uncertain as to what damages the assignee could recover, since the assignor would not have expended money on the remedying of undiscovered defects and would presumably have obtained market price for his property.'

What is the position with building defects? If the defects are patent at the time of sale and assignment of benefits, the purchase price received by the assignor will invariably reflect the diminution in market value and the assignment of benefits will be qualified accordingly so that it is the assignor who sues those responsible for the defective construction works. If the defects are latent then more than likely the property will have been transferred for its full market price. As the assignor has received full market price and will not in fact be expending any money on carrying out necessary repairs, has he suffered any loss which can be recovered by the assignee? These questions were considered by the Court of Appeal in the case of *Dawson* v. *Great Northern and City Railway Co.* (referred to in paragraph 3.29 above) in 1904 and some 78 years later by the House of Lords in *GUS Property Management Limited* v. *Littlewoods Mail Order Stores Limited*. **5.32**

Dawson

As stated, *Dawson's* case was concerned with a claim of a right to statutory compensation. The head note to the case states that it concerns a claim of a right to compensation and not a claim for damages for a wrongful act. This headnote is somewhat misleading. The Court of Appeal was troubled by the principle of law that a bare cause of action was not assignable and it is submitted that this was the reason for the court's emphasis on the right of statutory compensation and that their comments on the issue of damages are still helpful to formulating an answer to *Question 1* above. In *Dawson* both the freehold and leasehold interests were sold to D for market value and the vendor did not appear **5.33** to have suffered any loss. However the court held that the assignments of the benefit of the rights of compensation, which were made at the same time as the conveyance of the freehold, and the transfer of the leasehold entitled D, the assignee, to recover in respect of the assignors' rights of compensation against the original debtor, the railway company. Sterling LJ stated:

> 'It appears to us that the intention of this deed was to place the plaintiff precisely in the same position as regards the defendants with respect to the lands conveyed as was previously occupied by (the assignor) and in particular to transfer to the plaintiff the compensation for structural damage to the conveyed property.'

GUS Property Management

5.34   The facts of the *GUS Property Management* case concerned a building in Queen Street, Glasgow, owned by Rest Property Co. Limited which was damaged in late 1970/early 1971 by piling works carried out on neighbouring property for and on behalf of Littlewoods Mail Order Stores Limited. Serious structural damage was caused to the building owned by Rest. Rest was a wholly owned subsidiary of a holding company which in April 1972 adopted a policy of rationalising its property portfolio involving the transfer to a newly created, wholly owned subsidiary company, GUS Property Management Limited, the plaintiffs, of various properties, including the Queen Street building. Accordingly in March 1975 Rest conveyed its Queen Street building to GUS for a figure of £259,618 representing its book value. In June 1976 Rest assigned to GUS all of its claims arising out of the negligent building operations carried out on behalf of Littlewoods. Following the assignment GUS brought proceedings against Littlewoods claiming damages for delict (negligence) against Littlewoods, the consulting structural engineers (subsequently abandoned), the main contractors and the specialist piling sub-contractors. GUS claimed alternative heads of damage. Their first head of claim was for the sum of £350,000 representing the diminution in value of the building based upon the difference between the respective values of the building in a damaged and undamaged state at the time of the building operations. The second head of claim was for the sum of £55,450 in respect of the costs of repairing the damage to the building, which costs had been incurred by GUS not Rest, after the date of the conveyance of the building but before the date of the assignment. The main contractors and the sub-contractors contended that GUS's claims should be dismissed for the following reasons namely:

(1) GUS were really seeking to pursue a claim which Rest itself could have pursued at the date of the assignment.

(2) The only relevant loss which GUS could claim title to recover was loss suffered by Rest and recoverable by Rest at the date of the assignment.

(3) Accordingly the sums spent on repairs by GUS themselves were irrecoverable.

(4) The alternative claim for diminution in the building's value was irrelevant since the costs of the repairs, being all that was necessary to achieve compensation, represented the proper measure of loss.

(5) In any event, since the property had been transferred at book value without any regard for the fact that the building had been damaged, Rest had suffered no loss and accordingly had no claim to assign to GUS.

In the court of first instance in Scotland the Lord Ordinary rejected the defendants' submissions and the defendants appealed to the First Division, the appellate court in the Scottish system. The First Division reversed the decision of the Lord Ordinary and dismissed the claims brought by GUS against the main contractor and the piling sub-contractor. GUS appealed to the House of Lords. **5.35**

It will be apparent that the central issues in this case concerned the precise nature of an assignee's claim for damages against the original debtor. The First Division approached the issue from the correct starting point namely that GUS as assignees could only sue in respect of claims which were vested in Rest at the time of the assignment. That is to say the assignee stands in the shoes of the assignor. However the First Division went on to find that the assignor Rest, and consequently the assignee GUS could not have brought a claim at the date of the assignment for the costs of repair works as Rest did not carry out those works, nor had it incurred any expenditure nor any obligation in respect of those works. As for the alternative claim the First Division felt that it did not have to deal with this claim as its quantification substantially exceeded the claim for remedial works. Nevertheless the First Division commented that Rest would have had no claim in respect of the diminution in value because they had suffered no loss; the price which Rest received was the book value and would have been exactly the same even if the building had been in an undamaged state. The court considered that the position would be the same if a building with latent defects was sold at its market value in an undamaged state. The House of Lords, reversing the decision of the First Division, held: **5.36**

(1) The best measure of the loss sustained was likely to, but need not necessarily, be the difference between the price obtained in the sale of the property in its damaged condition and the price it would have fetched in an undamaged state.

(2) In this case the figure of price was of no relevance in estimating the plaintiffs' loss.

(3) The depreciation in value and the costs of reinstatement of the building were alternative approaches to estimating the damages, the appropriate measure emerging after the leading of evidence.

(4) The facts as to the costs of remedial works carried out by the plaintiffs' themselves might have evidential value for the purpose of arriving at an estimate of the loss suffered by the assignor Rest.

5.37 Lord Keith delivering the leading judgment in the House of Lords stated as follows:

'Where the property is disposed of in an arm's length transaction for the price which it is fairly worth in its damaged condition, the difference between that price and the price which it would have fetched in an undamaged condition is likely to be the best measure of the loss and damage suffered. It may happen that the owner of the property disposes of it otherwise than by such a transaction. He may, for example, alienate it gratuitously ... It is absurd to suggest that in such circumstances the claim to damages would disappear ... into some legal black hole so that the wrongdoer escaped scot free. There would be no agreed market price available to form an element in the computation of the loss and so some other means of measuring it would have to be applied, such as an estimate in the depreciation in value or the cost of repair. ...

It is undeniable on the pleadings that Rest suffered some loss through the defendants' operations. Its building was seriously damaged. How is the loss to be measured in money terms? One approach is to consider the extent to which the value of the building was depreciated as a result of the damage to it. Another is to assess the cost of repairs necessary to restore the building to the condition it was in before the defendants' operation. Both these approaches involve a process of estimating, an exercise familiar to courts of law. ... There is no doubt that the plaintiffs' pleadings were not drawn with that degree of accuracy which counsel might normally hope to achieve. The draftsman does not appear to have had in the forefront of his mind a sound grasp of the true legal position, namely that the plaintiffs are suing not for their own loss, but for that suffered by Rest. It would, however, not be right or just to dismiss the action by reason of this formal pleading defect, which is capable of being put right by a similar amendment without any prejudice being suffered by the defendants. The plaintiffs' averments about their own expenditure on repairs to the building are not open to any objections so far as they are averments of facts. They have relevance, in my opinion, as indicating the scale of expenditure which it is likely that Rest would have required to incur if they had continued to own the building. The facts averred may thus have evidential value for the purpose of arriving at an estimate of the loss suffered by Rest which is what the plaintiffs, as assignees of the claim, are in substance seeking to recover.'

The claims in this case were claims for damages arising from negligence. **5.38** It is submitted however that the House of Lords' decision, in so far as it relates to the measure of damages of an assignee, is equally applicable to a claim for damages for breach of contract. It follows that the answer to *Question 1* above must be that the sale by the assignor for full market value or the fact that the assignor does not carry out the remedial works are matters going to the issue of *measure of damages* and do not prevent an assignee from recovering its cost of remedial works or diminution in market value. Some commentators have expressed the view that it is essential to the recovery of damage by an assignee that the assignment of the benefits of a collateral warranty is contemporaneous with the conveyance or transfer of the property and that the assignment, as well as the property, should be expressly stated to be part of the transaction for which the purchase price is being paid (see for example Cartwright 'The Assignment of Collateral Warranties' CLJ 1990, Vol. 6 No. 1). The authors agree that such suggestions are very sensible practicable steps; however they are not necessary as a matter of law in the light of *GUS Property Management*.

Question 2

The answer to *Question 2* is concerned with the consequential losses. If A **5.39** enters into a collateral warranty with B, the first tenant, who then assigns his lease to C together with the benefit of the collateral warranty, in the event that remedial works are necessary to the building causing consequential losses, for example loss of business profits, is the original debtor A liable for B's consequential losses or C's consequential losses? It may well be of course that because of the nature of C's business those losses are much greater than the losses that would have been incurred by

B and *vice versa*. In the *GUS Property Management* case the plaintiffs' claim included a claim for consequential losses in respect of the loss of rental income for the period of the carrying out of the remedial work. The House of Lords did not specifically deal with this point; however it is suggested that in the light of the wording of the judgment of Lord Keith a claim for consequential losses suffered by the assignor (B in our example) would be recoverable but not the consequential losses of the assignee (C in our example).

5.40   Consequential losses were specifically dealt with in the *Dawson* case. The plaintiff's claim in respect of her freehold interest consisted of:

(1) £666 13s 4d in respect of structural damage, that is to say the amount which would have to be spent on the property in order to re-instate it in the condition in which it was before the defendants' works were executed.

(2) £700 damage to trade stock which was an estimate of the sum which would be sufficient to recoup to the plaintiff loss occasioned by her by disturbance of her drapery business carried on upon the property and by damage caused, or likely to be caused, to stock during the period occupied in the re-instatement of the building.

5.41   As regards claim 1 the court held that this claim was recoverable stating:

'[This sum] we understand to be the amount which would have to be spent on the property in order to reinstate it in the condition in which it was before the defendants' works were executed; and we are unable to see that there ought to be any difference in this amount whether the property was in the occupation of [the assignor] or of the plaintiff, or whether the proceedings were taken in [the assignor's name] or the plaintiffs. . . . We think, therefore, that so far as this item is concerned, the defendants have not had any greater burden imposed on them than they would have had to bear if the proceedings had actually been taken in [the assignor's] name.'

5.42   However the court rejected claim 2 stating:

'The amount has been arrived at on the assumption that the plaintiff was the person in occupation of the property, and it is contended that it ought to have been ascertained on the basis that [the assignor] was the occupier. In our opinion the plaintiff cannot . . . recover a greater amount of compensation than [the assignor] could have got. . . . It seems to us that, in these circumstances, [the assignor] could not recover any damage under the head of "damage to trade stock", and, for the reasons already given, neither can the plaintiff.'

It is suggested that the answer to *Question 2* above is that the assignee will not be entitled to recover his own consequential losses but should, subject to questions of proof, be entitled to recover the category or type of consequential losses which would have been recoverable by the assignor. It may well be however that the courts will use the rules as to measure of damages to restrict what otherwise might be unreasonable awards of damages to the assignee.   **5.43**

CONTRIBUTION AND APPORTIONMENT

Where, by reason of a breach of contract or tort, one or more parties are liable to the plaintiff for the *same damage*, the plaintiff is entitled to recover the whole of his loss against each defendant: *Cassell & Co.* v. *Broome*. Whilst the courts will make apportionments of loss as between the defendants (see paragraph 5.47 below) the courts will not apportion as between the defendants and the plaintiff. This rule is important in respect of collateral warranties for example if a building defect is caused by the negligent design of two consultants and only one of those consultants has given a collateral warranty then that consultant may be liable for the whole of the plaintiff's loss.   **5.44**

If, however, the parties' breaches have caused *different damage* then the defendant's liability will be restricted to the damage for which he was responsible: *Baker* v. *Willoughby*.   **5.45**

The Law Reform (Contributory Negligence) Act 1945 empowers a court to apportion damages as between the defendant and the plaintiff where the plaintiff's fault has contributed to the damage, for example where defective building works have been negligently overlooked by the architect and negligently overlooked by the employer's clerk of works. It would appear that, save for one exception, contributory negligence is confined to damages for tortious liability and not damages for breach of contract: *Basildon District Council* v. *J. E. Lesser (Properties) Ltd and Others*. The exception is confined to cases where the breach of contract is co-extensive with a tortious liability for negligence, the latter liability existing independently from the contract: *Forsikringsaktieselskapet Vesta* v. *Butcher and Others*.   **5.46**

Contribution

5.47   As between the defendants the court is empowered to apportion blame by
virtue of contribution awards under the Civil Liability (Contribution) Act
1978. Section 1(1) of the Act provides that 'any person liable in respect of
any damage suffered by another person may recover contribution from
any other person liable in respect of the same damage (whether jointly
with him or otherwise)'. It follows that the courts can apportion blame
between several defendants regardless of whether or not the defendants'
liability arises from different contracts or from a tortious liability. In
deciding the issue of liability section 1(4) provides that a *bona fide*
compromise entered into by the party from whom contribution is sought
shall be conclusive evidence of liability provided that the factual basis for
the claim can be established. Section 1(5) of the Act provides that any
judgment of a court shall be conclusive evidence of liability.

5.48   The essential elements are *liability* and *common damage*. For example
A is the consulting engineer for a project and B the specialist services
consultant. A enters into a collateral warranty with C but B does not. As
a consequence of negligence on the part of both A and B in the design of
the mechanical services, the heating installation is incapable of achieving
the required outputs. The benefit of the collateral warranty has been
assigned to D, the tenant in occupation. D brings proceedings against A
under the collateral warranty but has no right to bring proceedings
against B either in contract or in tort as D's loss is purely economic. In
these circumstances, whilst B is clearly blameworthy, A will have to pay
the whole of the damages to D and will have no right of contribution from
B. If however A and B's design faults had caused a heating boiler to
explode destroying the building, whereby B had a tortious liability to D,
then A would have rights of contribution against B.

5.49   Section 1(3) of the Act provides that contribution may be recovered
from someone who has 'ceased to be liable' thus preserving the rights of
contribution where the party from whom contribution is sought could
have pleaded a limitation defence or a settlement prior to the Act. The Act
therefore reverses the decisions of *Wimpey & Co. Limited* v. *BOAC* and
Harper v. *Gray and Walker*.

5.50   Section 2(1) provides that the courts are to assess contribution on the
basis of what is *just and equitable* having regard to the extent of the
person's responsibility for the damage in question. Further by section 2(3)
the court must give effect to any limitation of liability clause contained in
any relevant agreement.

5.51   In *Equitable Debenture Assets Corporation* v. *William Moss Group Limited* *and Others* the Official Referee gave separate apportionments of liability in respect of design and workmanship problems arising from defects in curtain walling. In respect of the design he apportioned 25% to the architect and 75% to the specialist sub-contractor. In respect of workmanship, he put 5% against the architect, 15% against the main contractor and 80% against the sub-contractor. Another example of an apportionment is the case of *Eames London Estates Limited and Others* v. *North Hertfordshire District Council and Others* where the apportionment in respect of defective foundations was $32\frac{1}{2}\%$ to the architect and $22\frac{1}{2}\%$ each to the local authority, the original developers and the specialist sub-contractor.

LIMITATION OF ACTION

Meaning

Limitation of action is a statutory remedy which prevents a plaintiff from bringing proceedings after the expiration of specified time limits. The philosophy of the remedy is that defendants should not suffer the prejudice of stale proceedings and that plaintiffs should be encouraged to avoid delay. This philosophy is of particular relevance to the construction industry which is well known for the longevity of its disputes. Limitation of action in respect of breaches of contract is governed by the Limitation Act 1980 which came into force on 1 May 1981. Limitation of action in respect of tortious liability is governed by the Limitation Act 1980 as amended by the Latent Damage Act 1986. Some commentators have considered that the Latent Damage Act also applies to breaches of contract where the contractual duty comprises skill and care, i.e. is analogous to the tort of negligence. It is the authors' view that this is not the effect of the Latent Damage Act by reason of the wording of sections 2 and 5 of that Act.

5.52

Limitation Periods

Section 5 of the Act provides that an action founded on simple contract (see paragraph 1.45 above) shall not be brought after the expiration of six years from the date on which the cause of action accrued. Section 8 provides that the limitation period in respect of a contract by deed (see paragraph 1.46 above) shall be 12 years. Care needs to be taken therefore in executing collateral warranties for if the original contractual obligation

5.53

is a simple contract and a party enters into a collateral warranty by deed, then the latter document will have extended the original obligations by a period of six years.

5.54   The Act does not interfere with limitation periods established by other statutes; for example under the Civil Liability (Contribution) Act 1978 the period of limitation is two years from the date on which the right to contribution accrues, usually the date of quantification, and under the Defective Premises Act 1972 the period of limitation is six years after the completion of the building works or completion of rectification works whichever is the later.

Calculation of the Limitation Periods

5.55   The starting point of the calculation is the date of accrual of the cause of action. In contract, the general rule is that the cause of action accrues at the date of the breach of contract (unlike tort when the cause of action accrues from the date that the innocent party suffers damage). Usually there will be little difficulty in identifying the date of the breach of contract. However, it is important to note that in a construction contract, if it is an entire contract, the date of accrual of the cause of action in respect of defective building works is not the date when those works were carried out by the contractor but the date of practical or substantial completion and possibly the expiration of the snagging period. Further, designers and design and build contractors may have a continuing contractual duty to check their design and correct errors during the period of construction: *Brickfield Properties* v. *Newton.* See also *Chelmsford District Council* v. *T. J. Evers and Others* where the writ was issued against a designer more than six years after the breach of contract but less than six years after the date of practical completion. The court refused to strike out the proceedings.

5.56   A cause of action cannot arise until there is a party who can sue and a party who can be sued: *Reeves* v. *Butcher.* It follows that, even though a collateral warranty does not seek to create any greater liability than the original contractual arrangement, the execution of the warranty may operate to extend the limitation period. For example, A, an architect designs a structure in June 1982 pursuant to a contract with his employer B which was entered into in 1981. As the development approaches substantial completion in December 1984 B sells the development to a tenant C and in January 1985, to facilitate this sale, A enters into a collateral warranty in favour of C. A's design is defective. Under A's original contract with B, which is a simple contract, the limitation period will have expired by June 1988. Under the collateral warranty with C, which is again a simple contract, the limitation period will not have expired until January 1991. If the collateral warranty were a deed, the limitation period would not expire until January 1997. Those giving collateral warranties therefore should be careful to abridge the limitation period. (See paragraph 5.58 below.)

Time stops running for the purposes of limitation upon the issue of a **5.57** writ of summons or the service of a notice to concur in the appointment of an arbitrator: section 34 of the act.

Deliberate concealment

Section 32 of the Act provides for a postponement of the commencement **5.58** of the running of the limitation period if there has been fraud or deliberate concealment by the wrong-doer or mistake. Both fraud and mistake are construed strictly and accordingly are of limited relevance. In contrast, deliberate concealment is concerned with the situation where any fact relevant to the plaintiff's rights of action has been deliberately concealed from him by the party committing the wrongful act. Section 32(2) provides that a deliberate commission of a breach of duty in circumstances in which it is unlikely to be discovered for some time amounts to deliberate concealment of the facts involved in that breach of duty. For example, mortar bridges permitting the transmission of damp from the outer leaf to the inner leaf of a brick cavity wall. Indeed, the authors were instructed in a case which involved the presence of cement bags in the cavity of the brickwork! It would appear, however, that the breach must be deliberate and not merely negligent: *William Hill Organisation Limited* v. *Bernard Sunley & Sons Limited* (a case concerned with the statutory predecessor of deliberate concealment, namely fraudulent concealment under the Limitation Act 1939). However, the fact that the employer has engaged a clerk of works for supervising the construction works will not necessarily prevent the employer from relying upon section 32: *London Borough of Lewisham* v. *Leslie & Co. Limited* (another case dealing with fraudulent concealment). Each case must be decided upon its own particular facts.

In *Gray and Others (the Special Trustees of the London Hospital)* v. *T. P. Bennett & Son, Oscar Faber and Others and McLaughlin & Harvey Limited* a hospital development had been completed in 1963. In 1979, there was evidence of a bulge in a panel of brickwork as a consequence of which structural investigations were commissioned. These investigations revealed setting out errors in the concrete panels, resulting in an unsatisfactory fit of the brick cladding and wholesale mutiliation of the concrete nibs in order to fit the brickwork. The employer brought proceedings against the contractors some 25 years after the construction of the development. The court distinguished the situation in *Gray* from that in *William Hill Organisation* and found that the breaches of contract in relation to the concrete panels and nibs had been deliberately concealed from the employer's supervisors, and therefore time for the purposes of limitation did not begin to run until the employer had discovered or could with reasonable diligence have discovered (section 32(1) of the Act) the concealment of the defective work; the earliest date for discovery was November 1979, when the employer noticed the bulge in the brickwork.

Abridgement of Limitation Periods

5.59   For reasons set out above, it is important that a party before entering into a collateral warranty considers whether or not the document will extend the limitation period beyond the period created by the principal contract. If there is such an extension, then this can be provided for in the collateral warranty by an express condition abridging the new limitation period to correspond with the original period. Such conditions are valid: *Atlantic Shipping and Trading Company* v. *Louis Dreyfus & Co*. This case concerned a charter party which provided for the reference of all disputes under the contract to arbitration and also had a clause which stated:

> 'Any claim must be made in writing and claimant's arbitrator appointed within three months of final discharge and where this provision is not complied with, the claim shall be deemed to be waived and absolutely barred.'

The court held that this clause was not open to objection on the ground that it ousted the jurisdiction of the court.

Acknowledgment and Part Payment

5.60   Section 29(5) of the Act provides that where there is a claim for a debt or other liquidated pecuniary demand and the person liable acknowledges the claim or makes any payment in respect of it, the claim is to be treated as having accrued on or before the date of the acknowledgment or payment. There can be successive acknowledgments or part payments, each one of which will give rise to a fresh calculation of the limitation period. However, once the limitation period has expired, an acknowledgment or part payment will not revive the claim.

An acknowledgment has to be in writing and signed by the person **5.61** making it, and has to be an admission of liability in respect of a debt or other liquidated amount or of a sum which is capable of being ascertained from extrinsic evidence. If these requirements are satisfied, there will be an acknowledgment even though the debtor in the same document makes a statement that he will never pay the debt: *Good* v. *Parry*. On the other hand, a statement that monies had been paid on account and that money might be due was considered not to be an acknowledgment for the purposes of the Limitation Act: *Kamouth* v. *Associated Electrical Industries Limited.*

It is not considered that the act of entering into a collateral warranty **5.62** will constitute an acknowledgment. However, it could if the appropriate words evincing an admission of liability were contained within the body of the document or in the recitals.

Chapter 6

Tenants, Purchasers and Funds

6.1 A funding institution, that is to say an organisation putting up money for a development, will have a dominant motive in mind when looking at collateral warranties: that is to obtain further protection in relation to the money that they are lending over and above the protection that they are already likely to have through a legal charge on the property and the other normal methods for securing money that has been lent. In a similar way, the purchaser of a freehold development will be seeking through a collateral warranty to give himself as much protection as possible in relation to the investment he has made: he will wish to have the right to make claims against those responsible for the design and construction of the development should something that they have done, or something that they have not done but should have done, cause the purchaser to expend money. A tenant of leasehold premises in a similar way is looking to protect himself from a liability to repair the development which he will normally have undertaken by entering into a lease. In respect of those issues, therefore, tenants, purchasers and funds have an interest in common: seeking to protect their own interests and pass as much risk as possible to the parties who have created the development — the architect, the engineer, the quantity surveyor, the contractor and, sometimes, the major (and on occasions all) sub-contractors. In order to understand the pressures felt by tenants, purchasers and funds when looking at the issues of collateral warranties, it is necessary to consider their positions separately in a little detail.

THE POSITION OF A TENANT

The Tenant's Problems

6.2 The law relating to landlord and tenant is a very complex area. It is, perhaps, therefore not surprising that there has been a great deal of litigation in relation to the meaning and effect of leases. Fortunately, in considering the impact of leases on collateral warranties, it is necessary to look at a relatively small aspect of landlord and tenant law: the obligation contained in a lease as to repair, known as the repairing covenant. It is essential in a lease to define whether the landlord or the tenant is to repair and in commercial leases, the obligation to repair is usually imposed on the tenant. However, whilst it is easy to say that the obligation to repair falls on a tenant, the nature of that obligation depends on the precise wording of the repairing covenant. Sometimes it is difficult for a lawyer to advise his client as to the precise meaning and effect of a repairing covenant in relation to particular circumstances that have arisen. However, some repairing covenants can require the re-building of the premises, for example:

'To repair and keep in repair the Premises and in addition when necessary to carry out all works of re-building, re-instatement and renewal of the Premises (whether in whole or in part) notwithstanding the cost of or reason for such works to the intent and effect that at all times during and at the end of the Term there shall be upon the land a high class building in good and substantial repair . . .'

At the other end of the scale, there can be provisos excluding from the **6.3** tenant's repairing obligations a liability to remedy latent defects. Such a provision is often sought by an ingoing tenant of a new development, but not often granted by the landlord (except perhaps when there is an over supply of buildings and a lack of willing tenants). Such a clause might be in this form:

'Provided that nothing in this lease shall be construed as obliging the Tenant to remedy any Defect of whose existence the Tenant has within the first . . . years of the Term notified the Landlord or any want of repair which is attributable to such Defect and which manifests itself within such period . . .'

Clearly such a clause would require a careful definition of 'Defect'. **6.4** However, it is between these two extremes that most leases will fall and a typical simple form for a repairing covenant given by the tenant might be:

'To repair the Premises and keep them in repair.'

What is the tenant's position if he has such a form of lease on a new **6.5** building and he discovers that there are serious design and construction faults? Whilst the answer to this question will depend on the particular circumstances that have arisen and the precise form of the wording of the repairing covenant in the lease, the tenant is likely to be in difficulties in arguing that he does not have an obligation to put right those serious defects in design and construction at his own expense. It is to be doubted whether a covenant to repair would be sufficient to force a tenant to completely rebuild. The legal position in relation to a latent defect in a new building causing a state of disrepair for the purposes of a repairing covenant is, however, something that gives rise to complicated legal issues. What is tolerably clear is that if there is a lack of repair caused by a latent defect, then that lack of repair falls to be dealt with under the repairing covenant. The more difficult question is whether the latent defect itself falls to be rectified by the tenant. If the only realistic way of effecting the relevant repairs is also to rectify the latent defect, that is likely to fall within the repairing covenant of the tenant: *Quick* v. *Taff Ely Borough Council*. On the other hand, in *Ravenseft Properties Ltd* v. *Davstone (Holdings) Limited*, a tenant was required to lay out substantial sums to remedy a latent defect. The court in that case, however, did not set out a principle that remedying latent defects would always fall within a tenant's repairing covenant but they did make it clear that a latent defect was not necessarily outside the repairing covenant:

> 'The true test is, as the cases show, that it is always a question of degree whether that which the tenant is asked to do can properly be described as a repair, or whether on the contrary it would involve giving back to the landlord a wholly different thing from that which he demised': Forbes J.

6.6 On the other hand, in another leading case, the tenant escaped liability for the cost of remedying a latent defect and the landlord had to bear the loss: *Brew Bros Ltd* v. *Snax (Ross) Limited*. In that case, the adjoining owner brought a claim in nuisance against the landlord of the premises: a wall was moving by reason of undermining of the foundations by a drain. The landlord joined the tenant in the proceedings claiming from him a full indemnity in respect of the adjoining owner's claim on the basis of the obligations arising under the repairing covenant in the lease. The lease was for 14 years and the defect arose just after the end of the first year of the lease. The Court of Appeal held that the repairing covenant included the drains and that both the landlord and the tenant were liable in nuisance to the adjoining owner. It was also held that the work required to make the premises safe was more than repair and was not therefore within the repairing covenant. As to what constituted 'repair', Sachs LJ said:

> 'It seems to me that the correct approach is to look at the particular building, look at the state which it is in *at the date of the lease*, look at the precise terms of the lease, and then come to a conclusion whether, on a fair interpretation of those terms in relation to that state, the requisite work can fairly be termed repair. However large the covenant it must not be looked at *in vacuo*.'

It is also clear that there must be a state of disrepair before any question **6.7** can arise as to whether it would be reasonable to remedy a design fault when doing the repair: *Quick* v. *Taff Ely Borough Council*. In *Post Office* v. *Aquarius Properties Ltd*, unusually, there was an inherent defect in the building that did not cause disrepair: defective retaining walls permitted flooding of a basement, ankle-deep. No damage had been caused to the building by the flooding: the building was in the same condition as when it was completed and let. As there was no damage, the Court of Appeal held that the tenants were under no obligation to the landlord to remedy the defect.

The application of the present law since *D & F Estates* is exemplified by **6.8** the case of *Ernst & Whinney* v. *Willard Engineering (Dagenham) Limited and Others*. E & W took an assignment of a lease, containing a full repairing covenant, for a commercial office building. Before they moved into the building, they discovered that the air conditioning ductwork was defective (it leaked and there were restrictions impairing performance). The total cost of the remedial works was put at approximately £1.5m. E & W had sought to argue that the ventilation system was defective and that the defects caused physical damage to the building which in turn had been caused by the failure of the defendants to carry out the work with reasonable skill and care. The judge took the view that he could not accept that argument because the building that E & W had leased, as their predecessors had leased it before them, was the building as built containing the ventilation system as it was, whether or not it was defective or whether its defects were apparent or latent. He went on to say that whether or not the system was deficient in terms of workmanship and performance was a matter that could only be measured by reference to the installation contracts, to which E & W were not parties. E & W failed in their claim.

It is therefore easy to see that a tenant on a repairing covenant under **6.9** a lease is taking on a risk, the full extent of which may, of itself, be the subject of some uncertainty. It cannot be a surprise, therefore, that tenants look for every means of reallocating that risk to someone else in so far as they are able to do so. Following the developments in the law of tort they cannot seek redress against any member of the team who designed and constructed the development, and with whom they have no contractual relationship: that contractual relationship is created by collateral warranties.

Parties giving Collateral Warranties to Tenants

6.10   The tenant will be looking for warranties that will provide a remedy for him in the event of defects in workmanship, materials and design. It follows that different considerations arise depending on the method of procurement of the project. If the building contract was JCT 80, under which the contractor will normally have no obligation as to design, then the tenant will look for collateral warranties from the contractor (in relation to the construction quality risks), the architect and the engineer (thereby covering the design element). Quantity surveyors are sometimes but not always asked to give collateral warranties to tenants and, in any event, they are not usually involved directly in either design or construction at least in the context of a collateral warranty. However, the position may be different if a quantity surveyor is involved in fitting out works for the tenant, particularly if those works are being executed under the main contract, as opposed to a separate contract with the tenants.

6.11   Where the project is on a design and build basis and the contractor has responsibility, therefore, for every aspect of the design and construction of the project, the position is different. Clearly the tenant will look for a collateral warranty with the contractor. The contractor will have himself employed an architect and an engineer to carry out the design for him. Should the tenant take collateral warranties from the architect and engineer in these circumstances? If the tenant's collateral warranty with the contractor has been carefully drafted, then the tenant, in the event of a breach of the warranty, can bring proceedings against the contractor and the contractor can, if he so wishes, join the architect and/or the engineer into those proceedings. In this way a chain of contracts and liability is set up.

6.12   On the other hand, it is not unknown for contractors to cease trading through insolvency — it is this fear that can cause tenants to seek collateral warranties from the architect and the engineer as well as the contractor on design and build projects. In circumstances where the contractor has become insolvent, the tenant who has a collateral warranty with the architect and the engineer will still be in a position to bring proceedings against them at least in relation to matters which fall within their duties — this is likely to be only design because there are few contractors who will appoint architects and engineers on design and build projects to carry out any supervisory duties in relation to the quality of the construction work itself.

Irrespective of the form of procurement of the project, it is becoming **6.13** more common for the major sub-contractors (mechanical and electrical services) to be asked to give warranties to tenants. There are several good reasons for adopting this approach. Firstly, main contractors are not liable to third parties for the torts of their sub-contractors, who are in law independent contractors — indeed this was one of the difficulties faced by the plaintiff in *D & F Estates*. On a project procured on a basis such as JCT 80, if the tenant has a collateral warranty with the main contractor, and a defect arises which is a breach of contract by the sub-contractor (which will also be likely to be a breach of the main contract), then in theory there is a chain of contracts setting up liability and the tenant should be able to recover from the main contractor, who in turn will recover from the sub-contractor. However, that chain of contracts could be broken by, for example, insolvency of the main contractor and caution suggests that a tenant might be well advised, therefore, to take warranties from the major sub-contractors. If some of those sub-contractors are nominated and have carried out design, they will have entered into Employer/Sub-Contractor warranty agreements in the form of NSC/2. One of the purposes of such a warranty is to create a contract under which the sub-contractor accepts responsibility for reasonable skill and care in his design, selection of goods and materials, and compliance with a performance specification — in other words, the employer has a remedy against the nominated sub-contractor under NSC/2 in respect of design problems. NSC/2 does have provision for assignment but it may be preferable for the tenant to incorporate a design warranty within the collateral warranty that he may take directly in relation to workmanship and material obligations.

The most difficult type of project from a tenant's point of view in **6.14** relation to collateral warranties is a construction management project. By 'construction management' is meant a project where the main contractor is only managing the project and does not enter into any of the trade contracts himself — those trade contracts are between the trade contractor and the developer. The effect of such an approach on collateral warranties for tenants is fairly horrendous. Unless the developer will give it (and he probably will not), the tenant does not have one person, such as a main contractor, to look to for his warranty — he therefore faces the prospect of obtaining warranties from the vast majority of the trade contractors. Unless the construction management project has been set up with this point in mind (and not all have) the tenant may face an impossible task. This is all the more so where the building is multi-tenanted and every tenant is looking for a warranty from every trade contractor for the important elements of the construction of the building.

6.15   On management construction projects (where the management contractor enters into the trade contracts himself), similar issues arise as to those set out above in relation to a project on, for example, JCT 80. However each and every management contract and construction management contract needs to be considered carefully by tenants in order to ascertain precisely what it is that they need, or, alternatively, where they are offered warranties from some parties, whether or not what they are being offered does in reality meet the risk that they are taking on under their lease.

What Tenants look for in Collateral Warranties

6.16   The tenant is looking for the designers to undertake an obligation to him in respect of design; that certain materials have not been specified for use in the building; that the designers have professional indemnity insurance and will maintain it; and that they can assign the benefit of the collateral warranty to third parties when they come to dispose of the lease. From the contractor, the tenant will be looking for a warranty that the contractor has carried out the construction of the work in accordance with his contract with the developer; that he has not used in the construction of the building certain specified materials; and that the benefit of the collateral warranty can be assigned to third parties.

6.17   From a design and build contractor, the tenant will want in addition a warranty in relation to design and that the contractor has and will maintain professional indemnity insurance.

6.18   From sub-contractors and trade contractors, the tenant will be looking for similar warranties and the tenant should be particularly careful to be certain to obtain a design warranty where it is appropriate from those sub-contractors and trade contractors. For example, whilst the engineer may design structural steelwork, it is the practice within the construction industry for the steelwork sub-contractor to design the joints in the steelwork.

THE POSITION OF A PURCHASER

The Purchaser's Problems

The fundamental problem faced by a purchaser of freehold property in the United Kingdom is the principle of *caveat emptor*: buyer beware. Put more simply, on a sale of freehold land and premises, it is up to the purchaser to find out whether or not the building has been built properly and designed adequately and whether or not it contains any latent or patent defects. It is not possible to imply a term into a contract for the sale of property to the effect that the property is free from defects. This fundamental principle is, therefore, a serious issue faced by every purchaser of real property. It is for this reason that purchasers invariably appoint surveyors/engineers/architects to carry out a survey and report on the state of the building prior to the purchaser becoming legally bound to purchase the building. However competently those surveys and reports are carried out, they cannot, inevitably, uncover extensive latent defects, although they can and do regularly uncover patent defects. Indeed, carrying out a survey at a level required to uncover some latent defects would be likely to be very expensive indeed: for example, inspecting brickwork cavities with fibre optics to check for bridging of the cavity and, more importantly, the presence or absence of brick ties. It is to be doubted whether most vendors of commercial property would permit such surveys to be carried out in any event. Clearly, therefore, the purchase of freehold property carries risk for the purchaser.

6.19

Interestingly, a verbal collateral contract was established between a vendor and a purchaser as long ago as 1901: *De Lassalle* v. *Guildford*, see paragraph 1.4 above.

6.20

However, the practice of conveyancing has moved on since 1901 and oral representations such as arose in the *De Lassalle* case are usually now expressly excluded by a contract term. Sales of property, both domestic and commercial, are now invariably governed by the Law Society's General Conditions of Sale. The 1984 revision of those conditions provided that the purchaser acknowledged in making the contract that he had not relied on any statement made to him save one made or confirmed in writing. However, the Standard Conditions of Sale 1990 (produced by the Law Society and the Solicitors' Law Stationery Society Limited and which are intended to supersede the Law Society's Conditions of Sale) are rather different in relation to such matters and provide that if a statement in the contract, or in the negotiations leading to it, is or was misleading or inaccurate due to an error or omission, a remedy is provided. So that for

6.21

example if there is a material difference between the description or value of the property as represented and as it is, the purchaser will be entitled to compensation. It follows that in order to obtain compensation under such a provision, it will be necessary to prove that a statement was made (whether oral or in writing) and that it was misleading or inaccurate due to an error or omission. The 1990 Standard Conditions of Sale do not make any attempt to negative reliance by a purchaser on oral statements not confirmed in writing. It remains to be seen whether or not these new conditions, if they become widely used in an unamended form, will make it easier for compensation to be obtained. Certainly, the balance on the 1990 Conditions is more in favour of a purchaser in this respect than the 1984 Conditions.

6.22   Many cases have arisen on the answers given to preliminary enquiries; difficult issues have arisen on the pre-1990 Conditions of Sale as to whether a claim could be brought under the Misrepresentation Act 1967 in respect of the loss suffered by the purchaser in reliance on the representation contained in the answer to the preliminary enquiries. That remains, at least in theory, a substantial weapon in the hands of the purchaser if he has asked appropriate questions and received informative replies. However, when the property market is booming, the answer to particularly searching preliminary enquiries is more likely to be 'the purchaser must make his own enquiries'. The purchaser then has to decide, on a commercial basis having made such enquiries as he can, whether or not he wishes to proceed with the purchase. Inevitably, in this respect, the vendor of property is in a better position than the purchaser.

Parties giving Collateral Warranties to Purchasers

6.23   The same parties are likely to be involved in giving collateral warranties to purchasers as were discussed under paragraphs 6.10 to 6.15 above in relation to tenants. However, where the development is sold during construction or well after completion, the alternatives of novation or assignment of benefits may be more appropriate than warranties (see 10.10. to 10.13).

What Purchasers are looking for in Collateral Warranties

6.24   Different considerations will arise depending on whether the purchaser is completing his purchase before or after completion of the building project. It is not at all uncommon for a freehold commercial property development to be sold during construction by one party to another (for example, between pension funds). Depending on the particular circumstances, a project in progress is probably best dealt with by way of novation agreements with all the members of the professional team and the contractor, whereby the purchaser stands in the shoes of the vendor, the project otherwise continuing without interruption. Each project of this kind needs careful consideration on its own particular facts and with careful consideration of the types of contracts being used and the method of procurement of the project. In these circumstances, it may be unnecessary for there to be collateral warranties for the simple reason that the purchaser will take over all the benefits of all the contracts of the professional team and the contractor.

Purchasers after completion of the building project will be looking for **6.25** all the same things as were discussed under paragraphs 6.16 to 6.18 above in relation to tenants. In addition, they are likely to be looking for a right to obtain copies of plans, drawings, specifications, calculations and similar documents in relation to the design and construction of the development. This inevitably involves considering issues of copyright which will have to be dealt with in the collateral warranty at the same time as obligations on the relevant parties to provide copies of the relevant documents. It is clearly important to a purchaser who may wish to carry out alterations to the development that he has a facility to obtain necessary information in relation to the construction of the development. If the purchaser is intending to let the building, he will be seeking to have the ability to provide to tenants collateral warranties from the designers and contractors.

THE POSITION OF THE FUNDING INSTITUTION

The Funding Institution's Problems

The funding institutions behind property developments come in many **6.26** forms: banks, merchant banks, insurance companies and the pension funds of large corporations. Before entering into a funding arrangement, funding institutions will naturally satisfy themselves that their involvement will be satisfactory from a financial point of view but also that, if things do not go according to plan, they are as well safeguarded as can be achieved. There will always be a lengthy agreement between the funding institution and the developer allocating the risk between those two parties and setting out each party's rights, duties and obligations. If the architect, the engineer or the contractor get into difficulties, then as far as the fund is concerned that will usually be a problem for the developer to resolve and he must take the financial risk. On the other hand, what is the funding institution's position if the developer gets into serious difficulties, for example a serious breach of the finance agreement with the funding institution involving non-payment of interest or the insolvency of the developer?

6.27   It is in the area of developer default that potential problems arise for a funding institution in funding a development. In order to safeguard the investment they have made up to that point, the fund will wish to have arrangements in place from the outset that enable the fund, in the event of serious default by the developer, to take over the project themselves, or possibly through a third party appointed by the fund, so that they can secure completion of the project with a minimum of extra expense, disruption and delay.

Parties giving Collateral Warranties to a Funding Institution

6.28   The parties giving collateral warranties to a funding institution will be substantially the same as those referred to under paragraph 6.10 above in relation to tenants. However, the fund will almost certainly wish to have a warranty from the quantity surveyor for the reasons set out in paragraphs 6.26 to 6.27 above and because they will wish to rely on the quantity surveyor in relation to his financial duties.

What Funding Institutions are looking for

6.29   Funding institutions are looking for at least the same matters that are dealt with above at paragraphs 6.10 to 6.15 in relation to tenants. Funds may well seek wider duties than just design, workmanship and materials. Sometimes they seek to have the same duties owed to them as are owed under the principal contract. The extent of such duties can cause unintended results in a warranty (see 8.19 and 9.41). In addition to those, funding institutions will be looking for rights that enable them to take over and complete the project, but only if they wish to do so, in the event of serious default by the developer. This will usually take the form in law of a novation agreement (see paragraphs 3.43 to 3.45). The effect will be, for example, that the client in the Architect's Appointment will no longer be the developer but will be the funding institution (or a third party appointed by the funding institution) but that that will be the only change in the Architect's Appointment. In other words, the client will have changed but none of the rights, obligations or duties between the client and architect will have changed. One client is simply substituted for another client.

The funding institution will also wish to have the right to use drawings, **6.30** specifications, calculations and the like produced by the design team so that the fund is free to make use of those documents for the purposes of the development. Again, this will involve dealing with the question of copyright.

OBLIGATION TO ENTER INTO COLLATERAL WARRANTIES

In an ideal world, every contract between every party of a development **6.31** project and agreements for lease/sale agreements would be entered into on the same day. This is, of course, unrealistic. However the fact that it cannot be done does raise some potential difficulties in relation to collateral·warranties. The architect is often the first person appointed by a developer when he has a project in mind. At that stage, although it is possible, it is unlikely that the developer has in mind a particular tenant or purchaser. The same situation will probably apply at the time of the appointment of the engineer and the contractor. The funding institution will usually come into the picture some time after the architect and before the contractor. Unless there is a pre-let, it is likely that the tenant will be the last party in time to become involved with the project. It is often the tenant or the purchaser that has the most concern about the need for and the contents of collateral warranties.

Situations often arise where tenants will not agree to enter into leases **6.32** unless and until they have collateral warranties in the form put forward by them (or at least agreed by them). By that stage, there will be no legal sanction on the architect, engineer and contractor to give collateral warranties unless such an obligation has been incorporated into their terms of engagement and the building contract.

Where no legal obligation to enter into collateral warranties has been **6.33** imposed by the terms of engagement and in the building contract, the fact that there is no legal obligation on the architect, engineer and contractor to give collateral warranties may lead to an outright refusal by those parties to enter into warranties. The only pressures that may arise are the usual commercial pressures — there will be no legal obligation. If the architect, engineer and contractor agree to discuss warranties, then there will inevitably be a long drawn out and expensive discussion as to the wording and extent of the warranties. This can result in a tenant withdrawing from the proposed transaction.

6.34   With this in mind, provisions are sometimes seen in terms of engagement of architects and engineers and in building contracts along the following lines:

> 'The Architect shall within 14 days of written notice from the Developer enter into a collateral warranty with any party proposing to enter into an agreement for lease of the Project in such form as that party shall reasonably require.'

6.35   Such provisions are likely to be ineffective in law. For an obligation of that type to be effective, the terms of the collateral warranty have to be certain. There is nothing certain where terms are not set out. Such a provision purporting to impose an obligation to enter into a collateral warranty where the terms are not certain will not give rise to a right on the part of the developer to apply to the court, for example, for an order of specific performance requiring the architect to enter into a collateral warranty. These clauses are akin to an agreement to agree which is of itself unenforceable. The better approach is to have a clause something like the following:

> 'The Architect shall within 14 days of a written notice or written notices from the Developer enter into a collateral warranty and/or collateral warranties with any party proposing to enter into an agreement for lease of the Project and/or any part of the Project in the form of the draft collateral warranty annexed hereto as Appendix A.'

6.36   Appendix A can contain the full wording of the proposed collateral warranty for that particular architect or engineer or contractor. Such an obligation is more likely to be enforceable by the courts. The inevitable difficulty that arises is that the collateral warranty has to be drafted and in place prior to the appointment of the architect, engineer or contractor as the case may be. At that time, the tenant/purchaser is unlikely to be known. Each and every tenant and purchaser has its own views about collateral warranties. It may be that the one that has been drafted and forms Appendix A in the example above is acceptable but, equally, it may not be acceptable to the tenant. However, at least in these circumstances the tenant can be offered, for certain, a warranty; the fact that the architect, engineer and contractor know that they have to give a warranty in that form often enables amendments needed by a particular tenant to be more readily agreed than if there was no draft warranty agreed from the outset.

Chapter 7

Insurance Implications

Very few professional practices of designers have sufficient resources **7.1** within their own organisations to meet anything other than minor claims brought against them in respect of professional negligence. It is for this reason that professional indemnity insurance is so important to the construction professions. Architects, engineers and quantity surveyors should simply not be asked to give collateral warranties which may result in professional indemnity insurance cover not being available to meet a claim brought under the provisions of that warranty: for such a situation to arise cannot be in the interests of the construction industry professions or the tenant, purchaser or fund to whom the warranty has been given. It follows that the dimension of professional indemnity insurance should never be forgotten by any party to a proposed collateral warranty.

PRINCIPLES OF PROFESSIONAL INDEMNITY INSURANCE

Proposal Form, Disclosure and Risk

Insurance policies are nothing more nor less than contracts and the usual **7.2** rules as to the formation of contracts apply to contracts of insurance. In practice, the usual procedure is for the person seeking insurance to complete a proposal form — each company has its own standard form. The person seeking insurance will sign the proposal form and by so signing he is offering to accept insurance on the basis of the proposal form and the insurers' standard conditions (to which reference will be made in the proposal form, although the terms will not usually be set out). Those terms will contain a provision to the effect that the answers given on the proposal form are incorporated into and are the basis of the insurance. As soon as the insurers have accepted the proposal, they will then issue a policy.

7.3   Contracts of insurance are different to other contracts in one important respect: they are based on the principle of the utmost good faith of the parties (known as the principle of *uberrimae fidei*). One of the aspects of this principle is that the purpose of a proposal form is to enable the insurer to assess the risk so that he can decide whether or not to accept the proposal and, if so, on what terms both as to the conditions and the premium. Clearly, insurers cannot make a proper evaluation of those matters unless full disclosure is made by the person seeking insurance of every matter which is relevant to the risk. When making a proposal to an insurer, it is necessary to disclose all facts that are material and not to make a statement that amounts to a misrepresentation of a material fact. Such non-disclosure or misrepresentation entitles the insurer to avoid the policy.

Non-disclosure

7.4   Disclosure means disclosing all facts that are material. Material facts are matters that would have affected the mind of a prudent insurer in deciding whether to take the risk and, if so, on what conditions and at what premium. This involves disclosing all material facts that are actually within the knowledge of the person seeking the insurance and this duty is not limited by what the person applying for the insurance thinks is relevant. It is clear that facts which show that a risk is not an ordinary risk but a greater risk than the ordinary are material for this purpose.

The Cover

7.5   The intention of a professional indemnity policy is to provide an indemnity in respect of the designer's legal liability for damages in respect of claims brought against the designer for breach of professional duty. The policy is therefore written on the basis that it covers claims made during the period of insurance. The period of insurance of professional indemnity policies is usually 12 months. It follows that a professional indemnity policy will cover claims made against the insured during that 12 month period. Indeed difficulties can arise where the full extent of a claim is not known during one period of insurance but becomes known during a subsequent period of insurance at a time when a different insurer is on risk: see for example *Thorman and Others* v. *New Hampshire Insurance Co. (UK) Limited and the Home Insurance Company.*

Renewal

Professional indemnity policies can only be renewed with the same **7.6**
insurer if the insurer consents. At renewal, most professional indemnity
insurers will require the completion of a fresh proposal form and the same
duty of the utmost good faith arises on the completion of a proposal form
for renewal for the simple reason that it is in reality a proposal for a new
insurance policy. In any event, a new insurer may be appointed on
renewal.

Generally

The matters set out above are very much a thumbnail sketch of some **7.7**
professional indemnity insurance principles. They are intended merely as
an introduction to the subject for the purposes of dealing with the
particular points that follow in relation to collateral warranties and
professional indemnity insurance.

DISCLOSURE OF COLLATERAL WARRANTIES

It is clear from the changes in the law of tort (Chapters 1 and 2) that in the **7.8**
absence of a collateral warranty, the tenant of a building put up by a
developer would be unlikely to succeed in a claim in negligence brought
against the architect or engineer or contractor of the developer in respect
of negligent design or construction. A collateral warranty entered into
between the tenant and the architect, on the other hand, would enable
that tenant to pursue his claim. It must follow that the existence of
collateral warranties is a material fact which must be disclosed to insurers
prior to professional indemnity insurance being taken out *and* at renewal.
After all, against the present background of the law of tort, the existence
of collateral warranties must be a matter which would affect the mind of
a prudent insurer when considering whether to take the risk and, if so, on
what conditions and at what premium (see 7.2 to 7.4 above).

Indeed, the fact that there are so many different forms of collateral **7.9**
warranty that have been signed and are being signed suggests that each
and every collateral warranty actually entered into should be produced to
insurers at renewal — at the very least a schedule of collateral warranties
should be appended to the proposal form with a statement that they will
be produced to insurers if insurers wish to see them. This is patently a
burdensome obligation both on the insured and on the insurer but given

the law on material disclosure in relation to professional indemnity policies, any other approach by the insured is dangerous for the simple reason that it may be that the policy could be avoided by the insurer if disclosure is not made of these material circumstances. The simple fact that many brokers (but not all), most insurers (but not all) and most underwriters are not set up to deal with a volume of work of this sort on renewal is irrelevant to the principle that full disclosure must be given of material circumstances.

7.10   Another day to day problem is this: should the insured, who is asked to sign a new collateral warranty, obtain his insurer's agreement before he agrees to execute a warranty? The strict answer to that question is that another warranty is another material circumstance that will have to be disclosed on renewal; if on renewal disclosure is made and the insurer refuses cover at that stage, the insured runs the risk of being uninsured in respect of any claims arising under or out of that particular warranty. It is for this reason that a cautious view should be taken and all proposed collateral warranties should be shown to insurers and their agreement to cover the risk obtained before the warranty is executed. There are two methods of avoiding this burden.

7.11   The first is where particular standard forms of warranty have been approved by insurers for general use by people insured with them, then it will not be necessary for insurer's permission to be obtained on each and every occasion. The RIBA, RICS Insurance Services (RICSIS) and ACE insurance schemes approved at the time of its issue the use of the standard form warranty to a company providing finance in the BPF Form CoWa/ F (see Appendix 1). However, no consultant should assume that this is the case or that it will be the case for ever in the future. Consultants should check on a routine basis with their own insurer what the position is at that particular time.

Some insurers, such as the Wren Insurance Association Limited, have produced their own forms of warranty but the rules of Wren provide that every warranty (even in its own standard form) must have express approval of insurers if it is to be covered. Furthermore, no-one should assume that because some insurers approve particular standard forms, other insurers will automatically approve the same warranties. That is not the case for the simple reason that each and every insurance policy is a separate contract between the insured and the insurer. Finally, the mere fact that the BPF Form of Agreement for Collateral Warranty CoWa/F to be given to a company providing finance is said to have been agreed 'after discussion with the Association of British Insurers' does not mean that every insurer who is a member of that Association will give cover in respect of a liability arising under a warranty in that standard form — whether or not cover is given is a matter between the particular insured and the particular insurer by way of agreement or by way of endorsement on the policy. Furthermore, such approval as is given to particular standard forms by particular insurers is usually limited to cover where the warranty is entered into in its *unamended* form — for example, the approval of CoWa/F by the RIBA, RICSIS and ACE insurance schemes is limited to that warranty, in its unamended form only.

Secondly, the difficulties of constantly referring warranties to insurers **7.12** for approval can be overcome to some extent by agreeing with the insurer a suitable policy amendment by way of an endorsement. This is dealt with in more detail in paragraphs 7.33 to 7.35 below.

PARTICULAR INSURANCE PROBLEMS

Operative Clause

The purpose of professional indemnity insurance is to give an indemnity **7.13** to the insured in respect of his loss — no more and no less than his loss: *Castellain* v. *Preston*. The policy of insurance achieves this effect by what has become known as the 'operative clause'. This is the clause of the policy that sets out the nature of the risk that the insurers are taking and in respect of which they agree to give the insured an indemnity for any sum which the insured has to pay as a result of the loss occurring. The operative clause is therefore of the utmost importance in deciding whether or not a particular loss is covered by a policy. Having looked at the operative clause to see if cover is provided, it is necessary to look at the other terms of the policy to see whether there are, for example, any exclusions that would mean the insurer was not liable on the indemnity otherwise provided in the operative clause.

In looking at collateral warranties and insurance, therefore, it is **7.14** essential to look first at the operative clause of the insurance policy. Unfortunately, there are innumerable different policy wordings produced by different insurance companies mutuals and underwriters and, further, they vary from profession to profession. It is therefore impractical in a book of this sort to cover all the possible policy wordings, although some points are made below in relation to a small number of policy wordings. The overriding principle must be for the insured to check that the operative clause of the policy does provide cover in respect of liability that may arise under collateral warranties.

7.15 Fairly typical policy wording for an operative clause is:

'The Insurer agrees to indemnify the Insured up to the limit specified in the Schedule in respect of any sum or sums which the Insured may become legally liable to pay as damages for breach of professional duty as a result of any claim or claims made upon the insured during the period of insurance arising out of the conduct of the practice described in the Schedule as a direct result of any negligent act, error or omission committed by the Insured in the said practice or business.'

7.16 The most important words in that clause in the context of collateral warranties are '. . . as a direct result of any negligent act, error or omission'. These words have given rise to a considerable amount of litigation in relation to particular circumstances that have arisen. Clearly, it is important in the context of collateral warranties to be reasonably confident that those words will provide cover in respect of a claim arising under a contract as well as a claim for negligence as a tort. In practice, insurers usually accept that a breach of a duty under a contract (whether express or implied) to exercise reasonable skill and care will fall within the matters giving rise to indemnity under the operative clause above. Some support for the view that the word 'negligent' does not also condition 'error or omission' was given by Webster J in *Wimpey Construction UK Limited* v. *D V Poole*:

'A professional indemnity policy does not necessarily cover only negligence. In my view I must give effect to the literal meaning of the primary insuring words and construe them as to include any omission or error without negligence, but not every loss caused by an omission or error is recoverable under the policy. In the first place, which is common ground, it must not be a deliberate error or omission.'

7.17 Another type of clause found in professional indemnity policies, but not often in those for the construction professions, is:

'The Company will indemnify the Insured in respect of claims made against the Insured and notified to the Company during any period of Insurance against civil liability incurred in connection with the conduct of the Business carried on by or on behalf of the Insured. . . .'

7.18 Such clauses as that above which give the indemnity in respect of 'civil liability' are least likely to give rise to difficulties in relation to collateral warranties. Civil liability is to be construed in contra distinction from criminal liability and will be likely therefore to include claims in the tort of negligence and claims under a contract as well as other forms of action, for example breach of statutory duty.

There are further policies around where the operative clause gives an **7.19** indemnity against 'liability at law for damages' in respect of claims arising out of the conduct of the business. On similar words, arguments have been advanced in cases that those words did not provide cover for claims arising from breach of an obligation to exercise reasonable skill and care under a contract. In other words, if that argument were right, there would be no cover for collateral warranties. The English case in which this point was argued is *M/S Aswan Engineering Establishment Co.* v. *Iron Trades Mutual Insurance Co. Limited* where the operative words giving rise to the indemnity under the policy were:

> '. . . against all sums which the Insured shall become liable at law to pay as damages and such sums for which liability in tort or under statute shall attach to some party or parties other than the Insured but for which liability is assumed by the Insured under indemnity clauses incorporated in contracts and/or agreements. . . .'

Hobhouse J said of the submission that this clause limited claims under **7.20** the policy to claims made against the insured in tort:

> '. . . the meaning of "liability at law" was to be ascertained by reference to the ordinary use of language. The court should not strain to put an artificial construction on the phrase, especially when it was the insurance company which was seeking to rely on the strained construction of one of their own standard forms. The meaning of the relevant words was plain. It was not restricted to liability in tort, especially having regard to the express reference to such liability in the second half of the clause.'

It should be noted, however, that a different conclusion has been reached **7.21** on this point in two Canadian cases, albeit they are not binding on the English Courts: *Canadian Indemnity Co.* v. *Andrews & George Co.; Dominion Bridge Co. Limited* v. *Toronto General Insurance Co.*

There are policies in existence, particularly for design and build **7.22** contractors, where the operative clause appears to limit cover expressly to 'negligence', for example:

'We ... agree to indemnify the Assured for any sum or sums which the Assured may become legally liable to pay ... as a direct result of negligence of on the part of the Assured in the conduct and execution of the professional activities and duties as herein defined.'

7.23 The word 'negligence' is capable of having a particular meaning in law and that meaning is limited to claims in tort. If the word 'negligence' were to have that meaning in this operative clause, then the consequences for the insured would be very serious indeed for he would have no cover in respect of claims arising out of a breach of duty under a contract — either a collateral warranty or his terms of appointment unless there were also negligence as a tort. In the case of collateral warranties, it is unlikely that there could be a claim in tort, otherwise there would be no need for the warranty in the first place. Although it is conceivable that a judge construing this type of policy wording would interpret it liberally to include a breach of a duty under a contract to take reasonable skill and care, it is a risk that the insured might be well advised to avoid. If possible, therefore, it may be best to avoid such policy wording or, if it cannot be avoided, then to have the word 'negligence' defined for the purposes of the policy so as to include breach of any common law duty to take reasonable care or exercise reasonable skill and/or breach of any obligation whether arising from express or implied terms of a contract or a statute or otherwise to take reasonable care or exercise reasonable skill.

Exclusions

7.24 Every professional indemnity policy contains a list of matters in respect of which the insurer will not be liable under the policy: these commonly include the excess, a claim brought about by dishonesty, fraud or criminal act, a claim brought outside a specified geographical area, libel and slander (unless there is a policy extension to cover these matters) and personal injuries caused to a third party unless they arise out of breach of professional duty. In the policies of particular professions, exclusions will be found which are peculiar to that profession: an example is a restriction as to cover for surveys and inspection unless the survey or inspection has been carried out by someone who has one of the specified qualifications. However, for the purposes of collateral warranties, there is one exclusion clause that can give rise to particular and serious problems. A typical clause is of this type:

'The giving by the Assured of any express warranty or guarantee which increases the Assured's liability but this exclusion shall not apply to liability which would have attached to the Assured in the absence of such express warranty or guarantee.'

or

'Any claim arising out of a specific liability assumed under a contract which increases the Insured's standard of care or measure of liability above that normally assumed under the Insured's usual contractual or implied conditions of engagement or service.'

It is clear that most collateral warranties will be caught by such an **7.25** exclusion clause in the policy. It follows that the policy would not give an indemnity in respect of claims made under collateral warranties in those circumstances. These exclusion clauses have been common form clauses in professional indemnity policies for a number of years and they came about because insurers became concerned that pressures were being placed on professional people to enter into contracts which provided for a higher level of duty than that to be expected of a reasonably competent person of that qualification holding themselves out as having those particular skills. The basis of professional indemnity insurance is that the standard of care to be expected from the professional man is of the type considered in *Bolam v. Friern Hospital Management Committee* (approved in *Whitehouse v. Jordan*): see 4.2. Clearly if a higher standard is to be expected of a professional man than that by a contractual arrangement, then it is a matter that insurers are entitled to have disclosed to them as a material fact. The type of exclusion clauses set out above are, then, important to insurers but they have the unfortunate side-effect of excluding from indemnity under the policy claims under collateral warranties.

It used to be argued (prior to the changes in the law of tort) that all a **7.26** simple collateral warranty did was to put in writing the duties that were in any event owed in tort — on that basis, these exclusion clauses were not so important. Now that the position has changed in tort, these exclusion clauses are of the utmost importance to people who want to rely on professional indemnity insurance either directly or indirectly: architects, engineers, quantity surveyors, contractors, funding institutions, tenants and purchasers. The only way to deal with these matters, where there is such an exclusion clause, is to agree an appropriate amendment to the exclusion clause with insurers or, alternatively, to have a policy endorsement so that collateral warranties are not caught by the exclusion, subject to the wording of the endorsement. This aspect is considered further in paragraphs 7.33 to 7.35 below.

OTHER MATTERS OF CONCERN TO INSURERS

Economic Loss/Consequential Loss

7.27   When there are defects in the construction or design of a building, the losses suffered by the occupier or owner can extend well beyond the direct costs of repair. There could be loss of production in a factory by the production having to be either stopped or disrupted while remedial works are carried out; there could be the costs of removal to other premises while remedial works are carried out and the costs of renting those other premises; there could be loss of profit; there could be a claim for the cost of management time spent by the occupier/owner in dealing with all the consequences of the defective design or construction. All of these matters might loosely be called economic or consequential loss. They can and often do exceed the direct cost of remedial works.

7.28   Insofar as the insurance policy is concerned cover is given in respect of 'damages' and that word in the context of an insurance policy means what the insured is legally liable to pay to the third party. It is for this reason that insurers are concerned as to the wording of collateral warranties in relation to consequential and economic loss. On the whole, insurers would prefer to see economic and consequential loss excluded in collateral warranties but this is of course an approach that does not appeal to purchasers, tenants and funding institutions. Further consideration is given in Chapter 5 as to what damages can be recovered under collateral warranties; there is further discussion of the possibilities of limiting liability in collateral warranties by their wording at paragraphs 8.9 to 8.14 and 8.72 to 8.77.

Assignment

7.29   The whole question of assignments of collateral warranties is a matter of great concern to insurers. They feel that unlimited assignments, say from outgoing tenants to incoming tenants every time the tenancy changes, will inevitably lead to a greater risk of claims. Whether this will prove to be the case when compared with the situation before the change in the law of tort remains to be seen. On the other hand, it must be the case that assignments increase the possibility of claims compared with a situation where assignments are not permitted or are restricted. It is because of the anticipation by insurers of this increased likelihood of claims that insurers will often insist on a restriction in a collateral warranty on rights of assignment. A typical requirement of insurers is considered in paragraph 7.34 below.

Fitness for Purpose and other Express Guarantees

It is the case that the operative clause of many insurance policies will only provide cover for 'negligent act error or omission' — this is the standard to be expected of a reasonably competent professional man carrying on that particular profession. In order to establish a liability on that sort of basis, it is necessary to look at what an ordinary competent person exercising that particular skill would do and to compare that with the actions of the person against whom the negligence has been alleged. In practice this is done by seeking the views of other persons exercising that particular skill and asking them whether the action that was taken by the person whose actions have been questioned are above or below the standard to be expected of an ordinary competent person carrying on that type of work. That is the function of expert witnesses and without such expert evidence, claims against professional people could not proceed at all: *Warboys* v. *Acme Investments*.   **7.30**

However, where there is an express or implied warranty that a design or piece of construction will be fit for a particular purpose, then the test in law is very different to that for professional negligence. The only questions that have to be asked are what was the purpose, and was the design or construction fit for that purpose? If it was not, then there is liability even where there has been no negligence. Similar arguments arise in relation to guarantees of performance whether in relation to design or construction.   **7.31**

Insurance will not extend cover for such fitness for purpose obligations or guarantees. Given that position, a person who enters into such obligations is likely to find themselves without the benefit of their insurance cover.   **7.32**

POLICY ENDORSEMENT FOR COLLATERAL WARRANTIES

Some professional indemnity insurers who understand the difficulties for designers and contractors created by the commercial need to give collateral warranties have been prepared to agree a policy wording which is of great assistance. The aim of such policy wording is to set out in the   **7.33** policy, or by way of an endorsement, the circumstances in which insurers will extend cover for collateral warranties; on that basis, the designer or contractor can then decide, when faced with a proposed collateral warranty, whether or not that particular warranty falls within the cover provided by the insurer or whether the warranty will have to be put to the insurer to see whether cover can be given by the insurer. This has several advantages to the designer/contractor and the insurer: many collateral warranties can be signed without the need to bother brokers/insurers with the wording. In those cases, the designer/contractor can have some confidence that his policy will cover those particular warranties. If proposed warranties are put forward which do not fall within the insurer's permitted restrictions, then the designer/contractor is in a good negotiating position with the third party because he can use the lack of insurance cover if he were to sign the proposed warranty as an entirely proper reason for seeking amendments to it. Finally, the insurer will not be constantly bombarded with requests for approval of non-standard warranties and the number that have to be put to him for approval will be considerably reduced.

7.34   A typical form of wording that insurers may be prepared to agree so as to overcome the difficulties created by the sorts of exclusion provisions referred to at paragraph 7.24 to 7.26 above is:

'Notwithstanding Exclusion "X", indemnity provided by this policy shall apply to Collateral Warranties or similar agreements provided by the Insured but only in so far as the benefits of such Warranties are not greater or longer lasting than those given to the party with whom the Insured originally contracted and subject to the following exclusions, unless specifically otherwise agreed by the Company:

(i)   acceptance of or guarantee of fitness for purpose where this appears as any express term
(ii) any express guarantee including any relating to the period of a project
(iii) any express contractual penalty
(iv) any acceptance of liability for liquidated damages
(v)   any assignment of a collateral warranty or similar agreement to:

(a) more than two parties in respect of assignments to funders, financiers and bankers
(b) more than one party in respect of assignments to any other parties.

These exclusions shall not apply to liability which would have attached to the Insured in the absence of such collateral warranties or agreements.'

The difficulties of providing effectively in collateral warranties for restriction (as opposed to prohibition) of assignment are considered at paragraphs 3.36 to 3.42 above and 8.65 below.   **7.35**

PROBLEMS ON CHANGING INSURERS

All of the matters raised in this chapter need to be carefully considered both on renewal of professional indemnity insurance with the same insurers, and even more carefully on changing insurers. Alternatively, the broker may be the same and the insurance scheme may be the same but the insurance company with whom the scheme is placed or, indeed, the underwriters participating in the cover may be different. In such changing circumstances, the agreement of an insurer in one year of insurance to the wording of particular collateral warranties will not bind any insurer at the time when a claim is made in a different policy year, perhaps many years after the collateral warranty was entered into, unless full disclosure has been given as discussed above at paragraphs 7.8 to 7.12, or the wording of the warranty falls within the express terms of the professional indemnity policy as discussed at paragraphs 7.33 to 7.35 above. Brokers will often say that changing insurers can be dangerous in a professional indemnity context; in the age of collateral warranties, it is potentially even more dangerous.   **7.36**

Chapter 8

Typical Terms

8.1 Although there are very few widely accepted standard forms of collateral warranty, there are types of terms that often arise in collateral warranties. Those types of terms are considered here following a discussion of the general considerations that should apply to the drafting of collateral warranties.

GENERAL CONSIDERATIONS

The Principal Contract

8.2 A draftsman should always bear in mind that the intention of a collateral warranty is to create a contractual relationship that is collateral to the obligations created by the principal contract. That principal contract can take a great many different forms: Architect's Appointment, one of the several ACE Forms, the RICS Conditions of Engagement, the main construction contract and all the various forms of sub and trade contracts. The draftsman's guiding principle should be that the collateral warranty should not properly seek to impose any greater or more extensive obligations than those which are created under the principal contract; one exception to this principle is where a sub-contract between a contractor and a sub-contractor contains no obligations as to design whereas the sub-contractor is in fact designing. Clearly that design obligation should be the subject of a clause in a collateral warranty so as to create a contractual cause of action in the event of default in the design obligation.

8.3 The draftsman should be particularly careful to provide words and obligations in the collateral warranty that are entirely consistent with the words and obligations contained in the principal contract. Equally, he should avoid the temptation to put additional and onerous obligations in a collateral warranty that do not appear in the principal contract. To do so is likely to lead to extensive discussions about the precise wording of the warranty with each party who is asked to sign the warranty and, in extreme cases, to produce a situation on the project where the commercial reality of the building process is lost. This can sometimes happen where the warranties that are to be obtained are decided by the funding agreement between the developer and the fund prior to the involvement of the professional and construction teams for the project. The developer is keen to have his finance in place for without the finance there will be no project but it is a mistake in those discussions between the developer and the fund not to have regard to the commercial realities of the construction market place; the expectations of some funding agreements in relation to the collateral warranties that are to be obtained by the developer from the professional team, the main contractor and sub-contractors sometimes bear no relation to what is in fact achieveable subsequently with those other parties.

Purpose

The draftsman should never lose sight of the purpose of the collateral **8.4** warranty that he is drafting: simply put, it is to create a contractual relationship in circumstances where there are no duties in tort owed by one party to the other. It follows that the language of tort is inappropriate for use in a collateral warranty. Some draftsmen seek to create a tortious obligation by means of a contract term:

'The Architect hereby agrees that it owes to the Tenant a duty of care in tort.'

In circumstances where the House of Lords have decided that no such **8.5** duty is owed in tort, it is very difficult to see how the courts could be persuaded to accept that such a clause has the effect of creating any tortious duties. In relation to design, for example, the usual need will be to create a contractual obligation on the designer to use reasonable skill and care; the use of language appropriate to tort is wholly inappropriate in the setting up of contractual relationships. In any event, in the example given above, it is very difficult to attribute any meaning at all to a confirmation that a duty of care in tort is owed by one party to another.

Contract or Tort

The issue has often arisen as to whether or not there can be, as between **8.6** two parties, liability in both tort and contract in relation to the same set of facts. A further issue that arises is whether, if there is a contract, which may contain clauses excluding liability or limiting the consequences of liability, there can also be a duty in tort which is not subject to the same restrictions that are imposed in the contract in relation to liability under the contract. There has been a trend in recent cases towards looking only to the contract. For example, in *Tai Hing Cotton Mill Limited* v. *Liu Chong Hing Bank Limited* in the House of Lords, Lord Scarman said:

> 'Their Lordships do not believe that there is anything to the advantage of the law's development in searching for a liability in tort where the parties are in a contractual relationship. This is particularly so in a commercial relationship . . . their Lordships believe it to be correct in principle and necessary for the avoidance of confusion in the law to adhere to the contractual analysis.'

8.7 Although this point was not expressly before the House of Lords in the later case of *Murphy* v. *Brentwood District Council*, their Lordships seem to have had in mind that there could be liability in both tort and contract. In *Greater Nottingham Co-operative Society Limited* v. *Cementation Piling and Foundations Limited*, the Court of Appeal faced a contention by the employer that the existence of a collateral warranty in the old 'Grey Form' between them and the nominated sub-contractors gave rise to an argument that there was close proximity for the purposes of seeking to establish a duty of care in tort owed by the sub-contractor to the employer; the sub-contractor, on the other hand, argued that the existence of the collateral contract restricted liability in tort. In this case, it had been conceded by the employer that the collateral contract did not have provisions that were appropriate for the particular circumstances that had arisen and they had therefore pursued their claim in tort. The Court of Appeal adopted the analysis of Lord Scarman in *Tai Hing* set out above, and Purchas LJ said:

> '. . . In considering whether there should be a concurrent but more extensive liability in tort as between the two parties arising out of the execution of the contract, it is relevant to bear in mind:
>
> (a) that the parties had an actual opportunity to define their relationship by means of contract and took it; and
> (b) that the general contractual structure as between the Society, the main contractor and Cementation as well as the professional

advisers produced a channel of claim which was open to the Society.'

The court went on to find that the economic loss suffered by the employer **8.8** in this case was not recoverable in tort; one of the reasons for that finding was the fact that there was a collateral contract. It follows from this case that in drafting collateral warranties consideration needs to be given to the following points:

(1) The collateral warranty must deal with all the points that the parties feel are important; any omission in the collateral warranty in that respect is unlikely to be made good by a claim in tort (subject, of course, to the restrictions in any event imposed by the decision in *Murphy* v. *Brentwood District Council* but also bearing in mind the re-inforcement that was given in that case to the decision in *Hedley Byrne & Co. Limited* v. *Heller & Partners Limited*).

(2) The existence of a collateral warranty may well have the effect of making it difficult, if not impossible, for a court to find a duty in tort.

(3) A collateral warranty which contains a provision which expressly preserves rights in tort, if any, may have the effect of preventing the contract being considered when looking at whether or not there are co-extensive duties in tort.

Exclusion Clauses

The Unfair Contract Terms Act 1977 renders totally ineffective a term **8.9** that seeks to exclude or restrict liability for death or personal injury resulting from negligence (section 2(1)). 'Negligence' has a wide meaning for the purposes of the Act (section 1(1)) and includes:

- any obligation, arising from the express or implied terms of the contract, to take reasonable care or exercise reasonable skill in the performance of the contract
- any common law duty to take reasonable care or exercise reasonable skill (but not any stricter duty).

It follows from these provisions in the Act that it is not possible in a **8.10** collateral warranty to seek to exclude or restrict liability for death or personal injury resulting from, for example, failure by a designer to exercise reasonable skill and care in his design.

8.11   However, the provisions of the Act that deal with seeking to exclude or restrict liability for breach of contract (otherwise than for death or personal injury) only arise in circumstances where one of the parties deals as a consumer or on the other's written standard terms of business (section 3(1)). It is unlikely that one party will be dealing as a consumer in the context of collateral warranties on commercial buildings.

8.12   Difficult questions can sometimes arise as to whether or not a particular document is, for the purpose of the Act, 'the other's written standard terms of business'. Firstly, it might be argued that collateral warranties are not, in any event, written standard *terms of business* for they are ancillary or collateral to those terms which are contained in the principal contract. Secondly, many collateral warranties are the product of negotiation between the parties and such warranties will not have their exclusion clauses looked at under the provisions of the Act. Thirdly, what is the position where there is in a standard form of warranty, agreed by various trade and/or professional organisations, a clause, for example, putting a limit on the amount of money payable as damages for breach? There is some support in the case of *Walker* v. *Boyle* for the view that some standard forms of contract are 'industry' forms derived from negotiation by representative bodies of both parties and that, accordingly, such contracts would not be the other's standard terms of trading for the purposes of the Act.

8.13   If the draftsman is preparing a clause seeking to exclude or restrict liability for loss or damage, other than death or personal injury, by a contract term, and it falls to be considered under the Act, then for that clause to be effective in law it has to satisfy the requirement of reasonableness in the Act (section 2(2)). The Act states that test as:

> 'the term shall have been a fair and reasonable one to be included having regard to the circumstances which were, or ought reasonably to have been, known to or in the contemplation of the parties when the contract was made' (section 11(1)).

8.14   Where consideration is given at the drafting stage as to whether or not a restriction of liability to a specified sum of money is reasonable under the provisions of the Act, some help can be derived from section 11(4) of the Act which provides:

> 'Where by reference to a contract term or notice a person seeks to

restrict liability to a specified sum of money and the question arises whether the term or notice satisfies the requirement of reasonableness regard shall be had in particular to:

(a) the resources which he could expect to be available to him for the purposes of meeting the liability should it arise and
(b) how far it was open to him to cover himself by insurance.'

Contra Proferentem

It goes without saying that in drafting it is important to avoid ambiguities   **8.15** both within clauses and between clauses; the *contra proferentem* rule, which is discussed at 1.62, is a powerful tool for resolving ambiguities.

Limitation of Action

Consideration should always be given by the draftsman of a warranty to   **8.16** the position in relation to limitation. The law in relation to limitation of action is considered at 5.52 to 5.62.

Duty of Care Letters

Prior to the recent changes in the law of tort in *D & F Estates* and *Murphy*,   **8.17** it was not unusual to see letters written by, for example, an architect to a tenant, confirming that the architect owed to the tenant a duty of care. It is unlikely that such letters have or had any effect in creating contractual relationships, although each such document would have to be considered on its own words. Such letters can fail to be effective in law for lack of certainty and, in particular, there is unlikely to be any consideration given for the promise contained in the letter. Absence of consideration will be fatal to the formation of a contract. To avoid these problems, it is best to avoid the use of letters to try to create collateral warranties and to prepare an agreement in writing which can be executed by the parties under hand or by deed.

TYPICAL TERMS

One of the difficulties that all the parties to the construction process face   **8.18** is that there are no universally accepted standard forms of collateral warranty, save perhaps the BPF Form of Agreement for Collateral

Warranty; that standard form is only for use by an architect, engineer or quantity surveyor giving a warranty to a funding institution. Many of the problems in the agreement of warranties arise because there are a very large number of purpose-drafted forms of warranty in the market place — many funds, developers, retail stores and tenants have their own standard forms which reflect their own experience, fears, knowledge and lack of knowledge. It is therefore difficult and probably unproductive to look in detail at the precise wording of many of the clauses that appear in practice. In any event, the wording varies enormously from collateral warranty to collateral warranty. It is however possible to put into categories the types of term that are commonly found. A commentary on the BPF Form of Agreement for Collateral Warranty for use where a warranty is to be given to a company providing finance for a proposed development is to be found at paragraphs 9.26 to 9.80.

Design

8.19   An approach sometimes adopted is that a designer gives a warranty that he has and will perform his design agreement with his client in all respects in accordance with that design agreement. Although this has the benefit of being truly collateral to the principal contract, it may have the effect that the designer owes all the duties under his design agreement to the third party in addition to the client. This could result, for example, in a design and build contractor owing all the duties that he owes to the employer under the contract to a tenant including, by way of further example, an obligation as to completing on or before the completion date and an obligation to pay liquidated and ascertained damages. The effect, therefore, of such a wholesale incorporation of the principal contract into a warranty is likely to lead to wholly unintended results. However, some funds may want more than just a design warranty (see, for example, 9.41).

8.20   The better way of proceeding in relation to design is to repeat in the collateral warranty the clause, or part of the clause, that appears in the principal contract in relation to design. For example, the architect's duty in relation to the standard of his design is to be found at condition 3.1 of the Architect's Appointment which is a contract between the architect and his client:

'The Architect will exercise reasonable skill and care in conformity with the normal standards of the Architect's profession.'

8.21   In like manner, the ACE Conditions of Engagement, Agreement 3, which is the contract between the engineer and his client on a project where an architect has been appointed, provides:

'The Consulting Engineer shall exercise all reasonable skill, care and diligence in the discharge of the services agreed to be performed by him.'

The transposition of such obligations from the principal contract into the **8.22** collateral warranty is not difficult and should provide a basis for drafting and agreement of this clause of the collateral warranty to be given by architects and engineers with the minimum of difficulty and discussion.

Sub-contractors usually fall into one of two categories in relation to a **8.23** design warranty, depending on whether or not design is part of the obligations undertaken under the sub-contract.

If there is no design obligation in the sub-contract, then the sub- **8.24** contractor should be required to give a collateral warranty to the employer/developer in any event and irrespective as to whether he is to be asked to give warranties to other parties. If the sub-contractor is a nominated sub-contractor under the JCT 80, or a named sub-contractor under the JCT Intermediate Form of Building Contract IFC 84, or a works contractor under the JCT Management Contract, 1987, there are standard forms of employer/sub-contractor agreements and these are to be found respectively as NSC/2, ESA/1 and Works Contract/3. In respect of the standard of design to be expected of the sub-contractor/ trade contractor, all three agreements are in substantially the same form:

'The sub-contractor warrants that he has exercised, and will exercise, all reasonable skill and care in:

(1) the design of the Sub-Contract Works in so far as the Sub-Contract Works have been or will be designed by the Sub-contractor, and

(2) the selection of materials and goods for the Sub-Contract Works in so far as such materials and goods have been or will be selected by the Sub-Contractor, and

(3) the satisfaction of any performance specification or requirement in so far as such performance specification or requirement is included or referred to in the description of the Sub-Contract Works . . .'

Again, the incorporation of such obligations in warranties to parties **8.25** other than the employer/developer should not be too difficult or contentious.

8.26   Where the sub-contractor is carrying out his design as part of his obligations under the sub-contract, the sub-contract should deal with the question of design. In the absence of an express provision dealing with design, the liability of a designing contractor will be likely to be a fitness for purpose obligation (see 4.14 to 4.16). The difficulties in relation to fitness for purpose in the context of design are discussed below at 8.30. However, the Standard Form of Sub-Contract for domestic sub-contractors for use with the JCT Standard Form 'With Contractor's Design' (1981), DOM/2, contains an express provision in relation to design as follows:

> 'To the extent that the Sub-Contractor has designed the Sub-Contract Works (including any further design which the Sub-Contractor is to carry out as a result of a Variation required by the Contractor) the Sub-Contractor shall have in respect of any defect or insufficiency in such design the like liability to the Contractor, whether under statute or otherwise, as would an Architect or, as the case may be, other appropriate professional designer holding himself out as competent to take on work for such design who acting independently under a separate contract with the Contractor had supplied such design for or in connection with works to be carried out and completed by a building contractor not being the supplier of the design.'

8.27   The effect of such a provision (which is a mirror image of the provision in the JCT 'With Contractor's Design', 1981 Main Contract) is that the liability in respect of design is to be the same as that of an architect or other professional designer, namely, reasonable skill and care. The effect therefore of this clause is to prevent the implication of a term that would otherwise be implied to the effect that the design shall be fit for its purpose. Again, there is little difficulty in providing a similarly worded clause in a warranty to be given by a sub-contractor or main contractor to parties other than the original employer.

8.28   Difficulties are sometimes perceived where architects and engineers are engaged by contractors in relation to a design and build project of the contractor. Usually, the main contractor, as under the JCT 'With Contractor's Design', 1981, will have dealt with the question of design expressly in the contract between him and the employer/developer. In relation to design, therefore, the contractor is in a position, if he so agrees, to give warranties in relation to the design to the fund, a purchaser and tenants. The question therefore arises whether or not it is necessary for the architect and engineer of the contractor to give warranties also to the fund, purchaser and tenants. Such warranties will often be needed in relation to the fund and a purchaser (particularly a purchaser prior to completion of the building works) for the reasons set out below at 9.20, namely, the desire of a fund and a purchaser to protect their position if, for example, the contractor becomes insolvent. A similar issue arises in relation to design. Usually there will be a chain of contracts, providing a route for a claim: for example the tenant through the collateral warranty to the contractor and the contractor through his engagement of the architect to the architect. If the contractor becomes insolvent, then this chain of contracts will be broken and the tenant will have no right in tort against the architect. It is for this reason that warranties are sometimes sought from architects and engineers on design and build contracts.

A further important point that arises in relation to warranties from architects and engineers on design and build projects is the question of supervision or inspection of the contractor's work. It is in practice rare for architects and engineers to be employed by a contractor in that capacity on a design and build project. However, the Architect's Appointment and the ACE Conditions of Engagement both provide for the architect and engineer respectively to carry out certain inspection functions on the site. If the architect/engineer used their standard terms of engagement for their appointments on design and build projects and, subsequently collateral warranties are entered into with third parties by reference to those standard terms, a warranty as to inspection may be incorporated into the warranty, even though it was never intended by the contractor or the architect/engineer that inspection duties should be part of their function. It is therefore important for architects and engineers on design and build projects to be certain that they are not being fixed by the warranty with inspection duties to third parties in circumstances where they are not in fact carrying out such duties. **8.29**

Fitness for Purpose

It is common place to see obligations as to fitness for purpose in respect of design as provisions in draft warranties. Such provisions as to fitness for purpose in design are not usually sensible provisions to be incorporated in collateral warranties for the following reasons: **8.30**

(1) Designers' professional indemnity insurance policies, including those of design and build contractors, are written on the basis of

reasonable skill and care (see 7.5). Furthermore, professional indemnity policies usually exclude any liability assumed under a contract which increases the standard of care or level of liability above that which normally applies under the usual conditions of engagement (see 7.24 to 7.26). It is to the benefit of every party to a collateral warranty that its provisions do not prevent the designer from having recourse to his professional indemnity insurers should a claim arise. To put it another way, there is no commercial benefit in drafting and securing harsh provisions if, when liability is established, there is no money available to meet the liability.

(2) The basis of the appointment under the principal contract of designers is that of reasonable skill and care — designers warrant that they will use reasonable skill and care not that they guarantee to produce a particular result. There is a distinction to be drawn here between a term as to fitness for purpose that could be implied as a matter of fact to give effect to the actual intentions of the parties and a term to be implied in law (to give effect to the presumed intention of the parties). An express obligation as to reasonable skill and care does not exclude the former.

8.31 Sometimes there is an obligation found in collateral warranties to the effect that a contractor or sub-contractor warrants that materials will be fit for their purpose in so far as they have been or will be selected by the contractor/sub-contractor. Such a clause creates no particular difficulties where it is in relation to a contract for the supply of work and materials and the contract includes no design obligations: it is nothing more nor less than the position at common law (see for example *Young & Marten* v. *McManus Childs*). Where the term is to be included in a contract where there is design in addition to work and materials, then consideration must be given as to whether or not such an obligation seeks to impose a fitness for purpose obligation in relation to design; it is a vexed question in construction law as to where design stops and construction takes over. Where there is any doubt, it is likely to be commercially sensible for all the parties to make such a fitness for purpose obligation in relation to materials subject to a duty to use reasonable skill and care in the selection — that is the formula adopted in NSC/2, ESA/1 and Works Contract/3 referred to at 8.24 above.

Workmanship

8.32 Main contractors are often asked to warrant to third parties that they will carry out and complete the project in accordance with the building contract. The extent and scope of such warranties requires careful consideration. No contractor should be asked to give warranties to third parties that go beyond his obligations as to quality contained in the building contract. This warranty can be drafted by reference to the wording of the building contract in respect of quality. For example, a contractor giving a warranty to a tenant as to materials and workmanship in circumstances where the main contract is the Standard Form of Building Contract, JCT 80, might be asked to do so in the following terms (which are substantially based on clause 2.1 of JCT 80):

'The Contractor warrants to the tenant that he has and will use in the construction of the Works (as defined in the Building Contract) materials and workmanship of the quality and standards specified in and/or required under the Building Contract.'

The use of such a warranty avoids the dangers of giving a general **8.33** warranty to a tenant that the contractor will carry out and complete the works in accordance with the building contract; such an obligation may raise a great many questions as to which of the duties in the building contract the contractor is also intended to owe to the tenant.

Deleterious Materials

Every warranty contains a provision that certain materials will not be **8.34** specified for use and/or used in the construction of the project. It has to be doubted whether this is a sensible and logical way to specify what materials are to be used. The most usual place to specify the quality of the project and the quality of the materials is in the specification. If certain materials are not to be used on the project, then those materials can easily be incorporated into the specification as being materials that should not be used. If that were done, then there would be no need for a deleterious materials provision in collateral warranties.

There are other reasons for not having a list of particular deleterious **8.35** materials in a warranty. Firstly, such a list, being a list of materials that must not be used, can never, by definition, be a complete list. Secondly, there is the danger that, on particular wordings, anything that is not specified as deleterious can be used. Thirdly, and most importantly, there is no reason why the third party should not rely on the general and positive obligations created by warranties as to design and quality of materials and workmanship. Could it really be seriously suggested that an engineer would not be in breach of a duty to use reasonable skill and care in circumstances where he had specified the use of high alumina cement concrete for the structural beams in the roof of a swimming pool? The same point can be made in relation to all of the seven or eight most commonly stated deleterious materials. There is little point in trying to have a definitive list of specific negatives when there is available an overriding positive general duty to exercise reasonable skill and care.

8.36   However the presently perceived conventional wisdom (probably coupled with pressure from tenants and commercial conveyancing solicitors) is to have a deleterious materials provision. A warranty is usually required of the designer that he has not and will not specify certain materials. A common list these days is:

- high alumina cement concrete in load bearing structures
- calcium chloride used as an additive in the mixing of concrete for use in reinforced concrete
- crocidolite (which is blue asbestos)
- asbestos
- wood wool slabs used as permanent shuttering
- high alkali cement not conforming with certain British Standards when used with aggregates containing reactive silica (known in the building trade as 'concrete cancer')
- urea formaldehyde (a foam insulation product)
- sea washed or sea dredged aggregates for use in reinforced concrete (except insofar as they comply with BS 882: 1983; BS 812: 1975 and 1976; and the GLC Development and Material Bulletin no. 121 dated March 1979).

8.37   The list of so-called deleterious materials varies dramatically from project to project. Some of these lists are now becoming of an extreme length, and some of the items are so lacking in definition as to be unhelpful and pointless in a practical building sense let alone from the point of view of a lawyer trying to construe the meaning. In any case where the lawyer dealing with collateral warranties does not understand the technical aspects of these lists, he should seek advice from someone who does, probably an engineer. There are real dangers in lawyers incorporating long lists of materials given to them by their clients without careful consideration of both the technical and legal implications. Some examples of the problems are given below, although the list is not exhaustive:

- 'Woodcrete and chipcrete'

- 'Mundic blocks'
  It is probably the case that such blocks are limited to certain areas of the south west of England, but the so-called 'mundic reaction' between certain types of aggregates containing pyrites and cement may be a much wider problem in concrete itself than in mundic blocks.

- 'Calcium chloride'
  Such a chemical appearing on its own is bordering on the absurd; as a matter of strict chemical analysis, calcium chloride will almost certainly be present in a great many building products. What is intended, but is not effectively achieved, is to prohibit the use of calcium chloride as an additive in concrete for reinforced concrete.

- 'Materials containing fibres less than three microns in diameter and between five and 1000 microns in length'
  Such a definition is unhelpful in a legal document; it is intended to deal with a description of fibres that can be dangerous to health, for example, by inhalation leading to tumours similar to mesothelioma caused by asbestos.
  However, it is rather more sensible and more certain in a legal document to specify precisely which materials are regarded as being deleterious: if this definition is simply intended to catch asbestos or man-made mineral fibres or ceramic fibres or synthetic material fibres, then it will be better both from a legal and a practical point of view to spell out the precise deleterious materials that are not to be used. Such descriptions by size of fibres in microns will not be readily understood by a majority of people in the building industry and professions. Furthermore, a well-drafted clause should set out whether such fibres are banned only when they are loose and capable of being inhaled, or whether they are also banned when they are combined with, say, concrete or paint, or cement or other products which have the effect of stopping the fibres from being inhaled unless the product itself is disrupted in some way.

- 'Brick slips where they do not comply with BS: 5628, Part III and BS: 3291, Part V'
  This is a further example of a provision which is wholly unnecessary; any designer who provides for brick slips which do not comply with

current British Standards will be in breach of his overriding duty to exercise reasonable skill and care.

- 'Lead' or 'Lead in drinking water supply pipes'
  The former is inappropriate; there are very few buildings without lead flashings somewhere; in any event, the chemical element 'lead' will be present in many other building products, for example, capillary (soldered) joints on copper pipe runs. The latter is verging on the absurd. Does the draftsman really think that a building services engineer or a contrator will specify for use lead water supply pipes in the 1990s?

- 'Polyurethane foam or polyisocyanurate board'
  These are sometimes banned together with urea formaldehyde foam. All these materials are quite often listed as banned deleterious materials in warranties, but included on the same project expressly by the specification of the contract works: an example is composite metal cladding for roofs, walls and insulated doors, which will contain foam of one or another kind.

- 'Materials referred to as being hazardous to health and safety in *Hazardous Building Materials: A Guide to the Selection of Alternatives* edited by S.R. Curwell and C.G. March in the edition current at the date of this warranty'
  This is an attempt to incorporate wholesale into a warranty part of the contents of a book that runs to 139 pages.
  It is very much to be doubted whether the joint editors and the authors of *Hazardous Building Materials* intended their book to be used in this way. The book is a useful and helpful contribution to wider understanding of the hazards of certain building materials, the alternatives that are available, and the hazards, if any, of the possible alternatives. However, as a matter of law, it is almost impossible to give any useful legal construction to a clause in a contract that seeks to incorporate parts of a book when the parts that are incorporated are not readily identifiable, there being no precisely identified list in the book of 'materials referred to as being hazardous to health'; indeed, in one part of the book, general guidance is given as to potential hazards of materials when in position, when disturbed and in the environment in waste disposal, each ranked separately as to none reasonably foreseeable, slight/not yet quantified by research, moderate, and unacceptable.

Does the draft clause intend to treat as deleterious all materials save those where no hazard is reasonably foreseeable when judged against the material in position or when it is disturbed, or when it is disposed of? In short, this clause is a brave attempt to do something useful, but in practice is so lacking in definition and clarity that it should not be used by any party to a proposed warranty.

The green factor is now beginning to creep into collateral warranties with   **8.38** certain types of material, some of which are widely used in the construction industry, being branded as deleterious. Quite commonly found materials in this category are certain types of timber preservatives where it is clear that (leaving aside the manufacturing and distribution risks) the greatest risk occurs during and shortly after application of the preservative to the timber. Some of these chemicals are lindane, tributyltin oxide and pentachlorophenol.

Sweep up Provision

There is often a sweep up provision at the end of the deleterious materials   **8.39** clause; again from a drafting point of view, it is almost certainly more satisfactory to rely on the positive duty to exercise reasonable skill and care (see 8.20) than it is to have a sweep up clause as a negative obligation tucked on to the end of the deleterious materials provisions. However, a not untypical sweep up clause found in collateral warranties is:

'any other substance or method of use or incorporation which is or may reasonably be suspected to be unstable, inadequate, dangerous, combustible or otherwise unsuitable for building purposes or for the type of building or conditions which it was used or is the subject of statutory control or does not conform to British Standards.'

Such a clause is so widely drawn and inconsistent with the standard forms   **8.40** of building contract and sub-contract that it should not properly be incorporated into any warranty. In particular, such clauses are often in conflict with the provisions in contracts and sub-contracts that require contractors and sub-contractors to comply with instructions they are given under those contracts. Further, not every building material in use is the subject of a British Standard or Code of Practice. It does not necessarily follow that such materials are deleterious.

Inconsistency

Conflicts between deleterious materials clauses in collateral warranties   **8.41** and the provisions in the contracts to which the warranties are collateral are a potentially serious source of problem. What is to happen if a material banned in the collateral warranty is instructed to be used by the architect in circumstances where the contractor has an obligation under the main contract to comply with architect's instructions? For this reason, the banned deleterious materials should also appear in the specification for the project (in the main contract and the sub-contracts) and, for an abundance of caution, there should be a provision that the architect is not permitted to instruct the use of those deleterious materials.

8.42   Sometimes, there is an additional provision requiring the giver of the warranty to notify if materials which were not deleterious at the time of incorporation have become generally known to be deleterious prior to practical completion. Such an apparently desirable clause can also lead to difficulties. If such a notice is given by reason of the collateral warranty, what is to happen to the project that is actually being built? Is work to stop? Who is to bear the cost of replacing the materials, and the costs of the inevitable delay? All these points need to be borne in mind when looking both at warranties and at the contracts to which they are collateral. They also provide an added reason for keeping the deleterious materials clause in a warranty in short and precise form, avoiding a lack of particularity in the drafting and avoiding the incorporation of general obligations, which in turn may lead to difficulties in relation to the principal contract.

8.43   A final point is that whilst a designer, who is not also building, can warrant that he will not specify particular materials, he cannot properly warrant that he will see that the contractor does not incorporate such materials into the construction (even though the designer has not specified them).

8.44   On the other hand, of course, it may be possible for a design and build contractor to give such a warranty, but not where he does not have complete control over the specification. The contractor on the JCT 'With Contractor's Design' form should watch this point in relation to the contents of the 'Employer's Requirements' and the right of the employer to require a 'Change' (see 8.41).

Copyright

8.45   There will usually be provisions dealing with copyright in drawings and specifications and the like. Occasionally, a fund or tenant will seek to have the copyright themselves. Such a provision is in direct conflict with the provisions of the standard form conditions of engagement (including, for example, the Architect's Appointment and the ACE Conditions of Engagement). The simplest way to avoid this problem is for the copyright to remain with the designer but for a licence to be given in the warranty agreement in respect of the copyright but limited to the purposes of the development — it would not be sensible for a designer to give a licence in his intellectual property in the absence of such a limitation. For example, the Architect's Appointment reserves copyright, not only in all documents and drawings prepared by the architect, but also 'in any works executed from those documents and drawings'. Unless the wording of the warranty contains the appropriate limitation on the licence for use on the named project, the licensee may well be able to construct another building using the same design without payment of any further fee.

The copyright/licence provision needs careful consideration in any **8.46** event both as to its scope and its effect; this is another area of the drafting of collateral warranties where careful regard should be had to the copyright provisions of the contract to which the warranty is collateral. Sometimes, there is not only a provision in the warranty granting a licence in respect of copyright but also an indemnity provision indemnifying the fund/tenant in respect of any proceedings, damages, costs and expenses which he may incur by reason of the giver of the warranty infringing or being held to have infringed any patent rights. The giver of such an indemnity must check that he is not giving an indemnity in respect of something over which he has no control. For example, a wide ranging indemnity is given by the contractor to the employer under the JCT 'With Contractor's Design' contract at clause 9.1. The indemnity is given in respect of various matters including infringement of patent rights and infringement of copyrights. However, there is a savings provision in the same contract at clause 9.2 so that where the contractor has supplied and used patented articles or inventions and the like in compliance with the employer's instructions, royalties, damages and the like which the contractor has to pay to the owner of the patent are paid to the contractor by the employer. It is important for contractors to see that there is no undermining of a provision such as that in any collateral warranties that they give.

The Copyright, Designs and Patents Act 1988 created for the first time **8.47** moral rights in relation to copyright. Such rights are to be distinguished from the copyright author's economic rights. One of the moral rights created by the Act is to give to the designer of the building a right to be identified as the designer 'by appropriate means visible to persons entering or approaching the building . . . and the identification must . . . be clear and reasonably prominent' (section 77(7)(b) of the Act). A person does not infringe that right unless the right has been asserted by the author of the design (section 78 of the Act). It is therefore the case that if the building owner/fund/purchaser/tenant do not wish the designer of the building to have the right to be identified by a prominent notice on the building, then they will need a provision by which the designer waives his moral right under section 77(7)(b) but designers would be well advised not to consent to any waiver of moral rights which goes beyond the right to be identified on the building.

8.48   Sometimes the granting of the licence is made conditional on the designer having been paid his fees in full; there may be further provisions that the licensee is entitled to call for the originals and/or copies of any and all of the relevant documents. The scope of such an obligation needs to be carefully considered by designers and they will wish to have a provision that they are paid the cost of the provision of copy documents.

Insurance

8.49   In warranties given by a party who is carrying out design, there is often a requirement as to professional indemnity insurance; these provisions vary enormously in their content and effect. A typical term might require that professional indemnity insurance is in force at the date of the warranty, that the premiums have and will be paid, and that the insurance will be maintained into the future, sometimes without limit in time and sometimes with a limit.

8.50   These kinds of provisions give rise to very real practical and legal problems. The first is that which arises from the nature of the professional indemnity insurance market. That market changes from time to time both in the average levels of premium required and in terms of the capacity and, consequently, the amount of the insurance (the limit of indemnity). It should also be remembered that professional indemnity insurance is annually renewable and is made on the basis that it covers claims made during the period of insurance. Given the inevitable vagaries of the professional indemnity insurance market place and the annual basis of this type of insurance, it is very difficult for a designer to give warranties in relation to future insurance, even one or two years ahead, let alone six or twelve years ahead. It follows that in looking at these types of clauses, the designer should have in mind the difficulty of forecasting the professional indemnity market as to availability of insurance, the size of the market and its cost. What is the designer to do if, six years after entering into this kind of obligation, the cost of the insurance becomes prohibitive to him in the context of his business at that time?

8.51   Secondly, this type of provision assumes that the designer's firm or company will continue in business in the future in the same legal form. Where it is a firm, the partners who entered into the warranty (and who are liable for breaches) may cease to be partners and/or may retire or die. Where the designer is a company, it may cease to trade without going into liquidation or receivership, or, it may go into liquidation or receivership. These points should be borne in mind by parties seeking from designers onerous insurance provisions in warranties.

Thirdly, where there is a prohibition, as there is in some policies, preventing disclosure of the existence of insurance to a third party, insurers' express permission should be obtained to entering into such a warranty. Depending on the terms of the insurance, the disclosure of its existence can have the effect of rendering the policy void. Further difficulties arise in relation to these provisions on renewal and on changing insurers (see 7.36). **8.52**

Fourthly, what is to happen if the designer is in breach of his obligation to insure? Very few clauses contain any provision to deal with this problem. Clearly the designer would be in breach of contract under the warranty and that would give rise to a claim on the part of the other party for damages. Those damages are the sum of money that would put the beneficiary of the warranty in the position he would have been in but for the breach. That sum might be the amount of the premium for the professional indemnity insurance. It is, however, highly unlikely that a third party would be able to take out professional indemnity insurance on behalf of the designer in any event. It follows that the consequences of breach of a provision of this kind are not terribly helpful to the party in receipt of the benefit of the warranty. **8.53**

To try to deal with all these factors, a very unobjectionable provision might be for the designer to warrant that there is professional indemnity insurance in existence at the date of the warranty and that the premium has been paid; the designer could further warrant that he will use his best endeavours to obtain professional indemnity insurance in succeeding years provided that such insurance is available in the market place at reasonable rates; these provisions could be extended by the designer undertaking to notify the other party or parties in the event that he cannot obtain insurance at reasonable rates so that a decision can be made as to how best to protect the positions of the parties. This kind of provision is to be found in the BPF Warranty to a fund (see 9.63). **8.54**

Novation

Funds and purchasers will wish to try to secure their investment and in **8.55** particular they may wish to have the right to continue with the project if the employer/developer is unable to continue, perhaps through receivership or liquidation or a serious breach of the funding agreement or the agreement to purchase. In order to try to give effect to this concern, funds and purchasers often require a provision in the warranty that they have the right to take over the project. What this means in practice is that in each warranty with each member of the professional team and the main contractor, those parties will agree that if requested by the fund or purchaser they will enter into a direct agreement with the fund in respect of the completion of the design/construction for the project. In principle there is nothing objectionable to such an arrangement, but there are major points that need to be carefully considered in relation to each of the different types of provisions that appear in practice.

8.56   The first point is the approach to the drafting; on a proper view, this arrangement is a novation agreement whereby, for example, in the building contract, the contractor continues as before and the fund or purchaser as the case may be are substituted for the employer/developer on the basis that the fund/purchaser take on all the rights and obligations of the employer. In other words, there is a substitution of parties. This would mean, for example, that where the contractor had existing claims against the employer (say in respect of unpaid interim certificates), the fund/purchaser would be liable. It is not unusual for funds and purchasers to try to avoid a commitment as to past liabilities by providing that they shall only be liable on the agreement from the date of the novation. Clearly such a provision is not one that designers and contractors would wish to agree. This is, of course, a question of commercial balance but it does seem that if the fund/purchaser wish to have the right to take over the agreements, then they should also agree to be liable in respect of past matters.

8.57   What is to happen to the existing agreement between the designers and the developer, and the contractor and the developer, when the fund or purchaser exercise their right to step into the shoes of the developer? The terms of the principal contract should be carefully considered in this context: for example, if the employer/developer falls out because of liquidation, there are provisions in the building contract to deal with that matter. The consequences of entering into a fresh agreement with a third party is that such a step is likely to be a repudiatory breach of the principal contract. There must, therefore, be a provision to the effect that the designer/contractor will not be in breach of his principal contract if the fund/purchaser exercise their right to step into the shoes of the developer/employer.

It follows from the need for a novation agreement and the need to **8.58** avoid repudiatory breaches of the principal contract that these kind of arrangements need to be tri-partite, that is to say, in the case of a contractor, the contractor, the developer/employer and the fund/ purchaser should all be parties to the agreement. This facilitates the drafting of a novation provision whereby one party is substituted for another and also enables a provision to be incorporated so that if the right to novation is exercised, the designer/employer can agree that that will not be a breach of the principal contract.

The most satisfactory way to resolve these problems is for there to be **8.59** an obligation on the designer/contractor and on the employer/developer in a tri-partite warranty agreement to enter into a novation agreement if the fund so requests, by which novation, the fund will be substituted for the developer as if it had always been a party to the principal contract. In those circumstances, there will often be a release by the designer/ contractor of the obligations of the developer (the fund having assumed all those obligations), but there will not usually be a release of the obligations owed by the developer to the fund. As a matter of style, the novation agreement can either be contained within the body of the warranty or by appending a draft novation agreement to the warranty and by a clause in the warranty requiring the other parties to enter into that agreement within a period of time after notice in writing from the fund/purchaser. Consideration should be given in drafting the funding agreement and in the contract for purchase to deal with these issues.

Problems are often caused by the draftsman of these kind of provisions **8.60** who seek to amend in the warranty the terms of engagement of architects and engineers and the building contract with the contractor; their intention is to provide that the architect, the engineer and the contractor will not exercise the rights under their principal contracts to terminate or determine their engagement without giving prior notice to the fund and/ or developer. Sometimes a draftsman will seek to achieve this by simply deleting the termination and determination provisions in the principal contract. That is a very dangerous course of action which could lead, for example, to a contractor having a purported obligation to continue with the project for the liquidator of a developer. The simplest way to deal with these problems is to provide that the architect, engineer and contractor give to the fund/purchaser a copy of any notice of termination/ determination that they give to the developer/employer under or in respect of a breach of the principal contract. Convoluted provisions to deal with this issue are only likely to lead to lack of clarity and to uncertainty.

8.61   Finally, novation will usually be a wholly inappropriate provision to be included in a tenant warranty.

Assignment

8.62   This subject is so complex that it merits virtually the whole of the contents of Chapter 3 and it is necessary to have in mind when drafting all the issues considered in that chapter. The following is a summary of issues, which draftsmen may find helpful.

No Provision for Assignment

8.63   If there is no provision for assignment, then the benefit of a warranty is freely assignable without the consent of the other party. Such an assignment can be legal or equitable (see 3.5 to 3.21).

Prohibition on Assignment

8.64   Where there is in a contract an express prohibition on assignment, then such prohibition is likely to be effective in law. It follows that if there were a purported assignment in such cases, it is likely that that assignment would be invalid (see 3.40).

Restriction on Assignment

8.65   Problems can arise where a contract provision seeks to impose a limit on the number of assignments that can be made; this can occur where the draftsman is attempting to make the provision on assignment in the warranty consistent with the provisions in the professional indemnity insurance policy as to the cover under that policy where there are assignments of collateral warranties. For example, the insurance policy wording might restrict cover to two assignments. If that provision is simply incorporated into the warranty, then every time there is an assignment the assignee steps into the shoes of the assignor and he has the right to make two assignments — in other words, the counting never starts. The simplest practical legal solution to this difficulty is to provide in the warranty that assignment is prohibited save where the express consent in writing of the original giver of the warranty has been obtained. The provision could continue to recite that the giver of the warranty shall not withhold his consent where the assignment is a first or second assignment. Such a provision puts the control in the hands of the original contracting party who needs to have that control, namely, the insured in this example.

8.66   Sometimes, a provision is inserted into a warranty that the agreement is personal to the parties. If a contract is a personal contract then the benefits of that contract cannot be assigned either by a legal assignment or by an equitable assignment. The issue here is whether a contract becomes a personal contract simply because it is agreed between the parties that it is. There has to be doubt as to whether such a provision in warranties is effective (see 3.36). It may be preferable, therefore, to prohibit assignment by an express term rather than seeking to do it by this indirect method.

Where there is a sub-lease, the first tenant remains liable on his **8.67** covenants to the landlord notwithstanding the covenants of the sub-tenant. It may be the case that the original tenant has assigned the benefit of collateral warranties by absolute legal assignment to the sub-tenant. If the sub-tenant then defaults on his repairing obligations (say through bankruptcy), the original tenant will remain liable on his repairing covenant to the landlord but he will not have the benefit of the collateral warranties. It is for this reason that it is sometimes suggested by funds and developers that the givers of warranties should agree to enter into fresh warranties every time there is a new tenant of the same premises. The effect of that request is to ask the givers of the warranties to insure part of the commercial risk inherent in the operations of landlord and tenant. This new warranty approach is likely to be strongly resisted by all parties who are asked to give warranties; consideration might be given to a provision in the assignment requiring re-assignment of the benefit to the first tenant if certain circumstances were to arise. However, such a provision must be capable of being carried through in accordance with any restrictions on assignment that there may be in the warranty.

'Other Parties' Clause

The Civil Liability (Contribution) Act 1978 deals with contribution **8.68** between people liable in respect of any damage whether tort, breach of contract or otherwise (section 6(1)). Under the provisions of this Act where two or more people are in breach of separate contracts with the same third person producing the same damage, those two or more persons can claim contribution against each other. This is discussed in detail at 5.44 to 5.51.

Following the changes in the law of tort, contracts, in the form of **8.69** collateral warranties, are now being widely used to fill the gap left by that change. Concern has arisen amongst some parties, particularly the construction professions, that if they give warranties and other people do not, they will be liable to the third party for the full amount of the loss and unable to claim contribution from those other parties who did not sign collateral warranties — those other parties can never be liable in respect of the same damage, there being no remedy in tort. The givers of warranties have therefore sought methods to try to make sure that they are not liable for the full amount of the loss in circumstances where other parties ought properly to be liable at the same time. The creation of enforceable provisions to this effect is difficult.

8.70   The most effective provision is likely to be that a warranty does not come into force and effect at all, a condition precedent, unless certain named other parties have entered into collateral warranties with the same third party to the same effect. This is sometimes called the 'Three Musketecrs' clause — all for one, one for all. Not surprisingly, developers, funds, purchasers and tenants are not keen on these kind of provisions. Another method may be for the developer to give an undertaking in the warranty that he will obtain warranties in the same or very similar form from certain other named parties. This is the approach in the BPF warranty and it is discussed in detail at 9.70 to 9.75.

8.71   There has been debate in legal circles about the possibility of a so called 'net contribution clause'; what is envisaged here is that by a clause in the collateral warranty, the liability of the person giving the warranty would be only that which it would have been had the other parties who might have been liable been before the court and the court had made an apportionment under the Civil Liability (Contribution) Act 1978. It seems that there is no effective way of drafting such a clause because it inevitably involves consideration as to whether or not other parties would have been liable and if they would have been so liable, what apportionment would have been made by a judge. These issues will have to be considered against the background that they involve the proper legal construction of a clause in a contract, and, a claim made *under* that clause, not as damages for *breach* of contract (where different considerations will apply: see 9.70 to 9.74). Clearly the effectiveness of such a provision would be doubtful. Furthermore, any attempt to try to give a court power to make that apportionment in circumstances where the other parties were not before the court would fail: the courts are not given power to deal with apportionment between parties who are not in fact liable in respect of the same damage. A court cannot be given, by a contract term, a jurisdiction that it does not have.

Limiting Liability

8.72   The parties to a contract can in English law agree a term excluding liability or limiting the consequences of liability subject only to some statutory regulation. The most important statutory regulation is the Unfair Contract Terms Act 1977 and the basic principles relating to that Act are set out at 8.9 to 8.14. It is the case that the courts when construing exclusion clauses are more likely to find acceptable a clause that purports to limit the consequences of liability, rather than excluding it altogether (see for example *George Mitchell (Chesterhall) Limited* v. *Finney Lock Seeds Limited* and *Ailsa Craig Fishing Co. Limited* v. *Malvern Fishing Co. Limited and Another*). In drafting a clause seeking to impose a money limit on a claim for damages, section 11(4) of the Unfair Contract Terms Act is helpful (see 8.14). This provides that, in looking at whether a particular term is reasonable under the Act, it is open to the court to have regard to two things: firstly the resources which that party could expect to be available to him for the purposes of meeting the liability should it arise; secondly, how far it was open to him to cover himself by insurance. This section gives some support to the view that it might be reasonable to limit liability to the limit of indemnity in the professional indemnity policy (provided, of course, that that limit of indemnity was a reasonable limit having regard to the type of work being undertaken).

The whole question of consequential loss is a matter that gives rise to concern amongst the givers of warranties. It is one thing, they say, to be liable for the direct costs of remedial works following defective design or workmanship, it is another to have to pay all the other economic consequences that may flow from such a breach of contract. Those other consequences can include loss of rent, the costs of the tenant moving out of the building whilst repairs are carried out, the costs of disruption/loss of profit in their business and the costs of returning after the remedial works. Sometimes, these indirect costs can exceed the cost of the remedial works. **8.73**

In contract, the damages that can be recovered are those that would put the innocent party in the position he would have been in, in so far as money can do it, but for the breach of contract. The kind of damages that can be recovered are governed by the question of remoteness and this is discussed in detail at 5.7 to 5.13. In principle, the kind of indirect losses discussed in the previous paragraph may well be recoverable as damages for breach of a warranty obligation. Various attempts have been made, therefore, to try to restrict that potential liability. One method is to exclude economic and consequential loss. However, those words do not have a precise legal meaning for this purpose and could give rise to difficulty of interpretation on particular facts. There is now a tendency for draftsmen to provide expressly for what is covered, rather than seeking to **8.74** limit the scope of general damages. For example, an architect might seek to limit his liability under the warranty to the direct cost of remedying defective work, all other costs, losses, damages and expenses being excluded. Developers, funds, purchasers and tenants are not keen on these limitation provisions.

8.75   Again, in looking at limitation of liability, the principal contract should not be ignored in drafting the warranty. For example, the Standard Form of Building Contract, JCT 'With Contractor's Design', 1981, contains a provision limiting the liability of the contractor in respect of design (except where the contract is for the provision of dwellings where the Defective Premises Act 1972 prohibits excluding liability). Clause 2.5.3 of that contract limits the liability for 'loss of use, loss of profit or other consequential loss arising in respect of the liability of the Contractor' for an insufficiency in design to a sum which has to be stated in the Appendix to the contract. In the event that such a limitation is agreed under the main contract, then that limitation should properly follow through into the warranty.

8.76   It is common for professional indemnity policies to provide cover for collateral warranties by way of a policy endorsement 'but only in so far as the benefits of such warranties are not greater or longer lasting than those given to the party with whom the insured originally contracted' (see 7.34). It may well be, therefore, that such a provision should be included by the insured in any warranty that he gives. As to 'greater', the person giving the warranty should be careful to see that the warranty is truly collateral and does not contain any more wide ranging obligations than he has under his principal contract. Sometimes this problem is met with a clause that provides that the damages recoverable under the warranty shall not be any greater than if the receiver of the benefit of the warranty had been, jointly, a party to the principal contract. The wording needs to be drafted with great precision — use of the words 'greater liability' will usually not create sufficient clarity. Sometimes, a similar formula is adopted in relation to 'longer lasting' (see 8.79 below).

8.77   During the construction of a project there can be various agreements between, for example, the architect and the employer, and the contractor and the employer, in relation to, for example, remedial action in respect of defective work. The only remedy available to an architect under earlier editions of JCT 80 was to require removal from the site of defective works. Now, clause 8 of JCT 80 provides that, in certain circumstances, defective work can be remedied but left in place. That example leads to consideration in warranties of a provision that any agreement under the principal contract will also bind the beneficiary of the warranty. This problem can be met to a certain extent by making sure that the main operative clauses of the warranty are truly collateral but every giver of warranties should consider whether a wider clause should be expressly incorporated to cater for this potential problem.

Limitation of Action and Indemnity Provisions

The law in relation to limitation of action is considered at 5.52 to 5.62. In so far as drafting is concerned the following points are important. **8.78**

It is unlikely that simply entering into a collateral warranty constitutes an acknowledgement of liability for the purposes of limitation (see 5.60 to 5.62); however, it is also the case that there cannot be a breach of a collateral warranty until a date after the date of the warranty. If the giver of the benefit of the warranty wishes to impose through the warranty some agreement as to limitation of action then it is open to him to do so for the reason that limitation periods can be altered by agreement. Sometimes the giver of the benefit of the warranty will seek to have a clause incorporated which says that he will have 'no longer lasting liability' than that arising under the principal contract. It is thought that those words may not have the desired effect for the simple reason that they may not be sufficiently clear in relation to limitation. A party desiring to effect such a limitation, therefore, should expressly set out in the clause the limitation period by reference to the principal contract; in other words, whilst the warranty may have been entered into at a later date, and no cause of action can arise before that date, the date by which the breach is to be taken for the purposes of the limitation period (six or 12 years depending whether under hand or by deed) will be the same date that would arise on a breach of the principal contract. **8.79**

The givers of indemnities under collateral warranties should be aware of the consequence that they have on the limitation position, viz. that the cause of action does not accrue on an indemnity until the liability of the person seeking to be indemnified has been established: *County & District Properties* v. *Jenner*; *Green & Silley Weir* v. *British Railways Board*. This principle of law can have the effect of substantially extending the period of time after which proceedings can be successfully brought. **8.80**

Delay

On a building project, the building owner will usually be able to recover liquidated and ascertained damages from the contractor in the event that **8.81** he is late in completion, subject to any provisions for and the granting of any extensions of time. The whole issue of delay *vis à vis* the contractor is therefore dealt with in this way. On a design and build project, any delay in the design of the contractor (or the people to whom he sub-lets the design) will also be the subject of liquidated damages and extension of time machinery. However, where the main contractor is not designing, such as where JCT 80 is used, then it is the case that the architect, the engineer and any designing nominated sub-contractors can cause delay to the main contract. In each of those cases, the main contractor will usually be entitled to an extension of time in respect of the delay caused by the issue to him of late information. In these circumstances, the standard form of nominated sub-contractor/employer warranty, NSC/2, has certain provisions.

The first is that the sub-contractor agrees to supply the architect with information in due time so that the architect will not be delayed in issuing that information to the main contractor such that the main contractor may have a valid claim to an extension of time for completion (clause 25.4.6 of JCT 80) or a valid claim for direct loss and/or expense (clause 26.2.1 of JCT 80). The effect of such a warranty is that, whilst the employer will not be able to obtain liquidated damages from the main contractor by reason of the extension of time that has to be granted by the architect, the nominated sub-contractor renders himself liable to that claim from the employer together with any claim for direct loss and/or expense that the employer has to pay to the main contractor as a result of the sub-contractor's late information.

8.82   Secondly, under NSC/2 the sub-contractor warrants that he will so perform his obligations under the sub-contract that the architect will not be under a duty to issue an instruction to the main contractor to determine the employment of the sub-contractor. Again, breach of this warranty to the employer may render the sub-contractor liable to costs incurred by the employer under the main contract in circumstances where the sub-contract has been determined. Thirdly, the sub-contractor warrants under NSC/2 that the contractor will not become entitled to an extension of time for completion of the main contract works by reason of delay on the part of the nominated sub-contractor. Again, breach of this obligation will entitle the employer to recover main contract liquidated damages from the sub-contractor.

8.83   Consideration should be given as to whether clauses of this type should be incorporated in collateral warranties and/or in the principal contracts (for example the architect and engineer).

Disputes Resolution Procedure

Most collateral warranties are silent on the procedure to be adopted in the **8.84** event of a dispute. On the other hand, most of the contracts to which they are collateral contain arbitration provisions — see for example the Architect's Appointment, ACE Conditions of Engagement, RICS Conditions of Engagement, all the JCT contracts and sub-contracts.

In relation to a tenant's claim under a warranty at a time after **8.85** completion, this will give rise to no difficulties and the dispute can simply be heard in the court. However, in relation to a dispute that arises under a warranty at a time prior to the final certificate under the building contract there are many potential difficulties. For example, procedural problems will be created where there is a dispute under the main contract between the contractor and the employer as to, say, an employer's set-off of liquidated damages in circumstances where the contractor believes he is entitled to an extension of time. At the same time, the funder exercises his right to require novation, believing the employer to be in material breach of the funding agreement: the building dispute will be the subject of an arbitration clause and the dispute about the right to novate will be likely to be subject to the jurisdiction of the courts. In both cases, the resolution of the dispute turns upon the same facts and the same evidence and there is a clear risk of inconsistent findings between the court and the arbitration. There are three ways to deal with this matter:

(1) The first is to make no provision for disputes resolution procedure in the warranty and for all the parties to keep their fingers crossed.

(2) The second is to have an arbitration clause in the warranty together with multi-party machinery both in the warranty and in all the relevant principal contracts but limited to disputes on issues which are substantially the same as or connected with issues raised in the related dispute under the other contract. Such a multi-partite arbitration clause should give the arbitrator power to make directions and all necessary awards in the same manner as if the procedure of the High Court as to joining one or more defendants or joining co-defendants or third parties was available to the parties and to him. Clearly, such multi-partite machinery will not usually be relevant in warranties to tenants but may well be relevant to warranties given to the fund and the purchaser so as to avoid the risk of multiplicity of proceedings.

(3) The third method is to have a provision in the warranty that English law applies and the English courts shall have jurisdiction.

This does not overcome the risk of multiplicity of proceedings where there are arbitration clauses in the principal contracts.

8.86 Sometimes, developers and funds will seek to overcome the risk of multiplicity by arranging for the arbitration clause in the building contract (and in the other principal contracts) to be deleted and, in their place, an English law and English courts clause is inserted. This is a potentially dangerous route. Under JCT 80, for example, an arbitrator has the power to open up, review and revise any certificate, opinion, decision, requirement or notice given during the course of the works. It is to be doubted that the court has such power, whether litigation is commenced with an arbitration clause in the contract or where it has been deleted (*Derek Crouch Construction Limited* v. *North West Regional Health Authority*). Further, a clause inserted in the building contract purporting to give such a power to the court is not likely to be effective: the parties by a contract cannot give to the court a jurisdiction that the court does not have.

8.87 Finally, it is inappropriate to seek, as some draftsmen do, to delete the arbitration machinery in the principal contract by a clause in the collateral warranty. The proper place for amendments to the principal contract is in the principal contract and not in the contract that is collateral to it.

Notice

8.88 It is useful to include a provision as to how notices are to be given in the procedural sense: by post, recorded delivery, registered post, by hand, by fax, by telex; at what address and when service is to be taken as having taken place.

Under Hand/By Deed

8.89 Although strictly not a term of the contract, the basis upon which the execution of the collateral warranty is to be carried out is important for two reasons. Firstly, where the warranty is to be executed under hand, then the draftsman must provide for consideration (see 1.42). Secondly, warranties under hand are subject to a six year period of limitation of action whereas contracts executed as deeds are subject to a 12 year period of limitation of action (see 5.53). The receivers of the benefits of warranties will usually be looking for 12 years, and, therefore, deeds. As to the new provisions for execution see 1.46 to 1.48.

CONTRACTORS AND SUB-CONTRACTORS

A question arises as to whether or not there should be a provision in sub-contracts in relation to collateral warranties entered into by the main contractor. There are two important points here. The first is whether or not the limitation periods are different under the sub-contract and the main contractor's collateral warranty — the danger point for the main contractor is expiry of the limitation period under the sub-contract at a time when the limitation period under the collateral warranty has not expired. Secondly, it should be made clear in the sub-contract that any damage that the main contractor suffers under a collateral warranty are damages in the legal sense that are recoverable under the sub-contract.   **8.90**

As to the limitation point, a useful device is to have an indemnity in the sub-contract. The cause of action does not accrue on an indemnity until the liability of the person seeking to be indemnified, the main contractor, has been established. The second point, foreseeability of damage, can be dealt with by a provision in the contract. Putting these two points together, the type of provision that main contractors might wish to incorporate in their sub-contracts is: **8.91**

'The Sub-contractor hereby acknowledges that any breach by him of the Sub-contract may result in the Contractor committing breaches of and becoming liable in damages under the Main Contract and other contracts made by him in connection with the Main Works (including but not so as to derogate from the generality of the foregoing collateral contracts with funding institutions, tenants and/or purchasers of the Main Works and/or any part thereof) and may occasion further loss or expense to the Contractor in connection with the Main Works and/or further or otherwise and all such damages, loss and expense are hereby agreed to be within the contemplation of the parties as being probable results of any such breach by the Sub-contractor. The Sub-contractor shall indemnify and save harmless the Contractor from and against all such damages, loss and expense as aforesaid.'

Chapter 9

Practical Considerations

9.1   A great many practical issues are raised on a regular basis by those involved in collateral warranties. Does a warranty have to be given at all? What the tenant wants is unfair; who is to pay the legal costs of negotiation of non-standard warranties? Do warranties have to be signed by the parties? How should the negotiations be conducted and how should professional indemnity insurers be involved? Should different terms be in the warranties depending on whether they are being given to funds, purchasers or tenants? The aim of this chapter is to look at those matters and then to provide a commentary on the Form of Agreement for Collateral Warranty for use where a warranty is to be given to a company providing finance for a proposed development prepared and approved for use by the British Property Federation (BPF), the Association of Consulting Engineers (ACE), the Royal Institute of British Architects (RIBA) and the Royal Institution of Chartered Surveyors (RICS), known as CoWa/F.

DOES A WARRANTY HAVE TO BE GIVEN?

9.2   A warranty does not have to be given by anyone unless they are under a binding and certain contractual obligation so to do, subject, of course, to the commercial pressure that can be brought to bear. It follows that in the absence of such a binding and certain contractual obligation, such as that at 6.35, no warranty whatsoever has to be given. This is particularly so where, after practical completion of a development, proposed tenants seek warranties from the architect, the engineer and the contractor. This gives rise to points which are the opposite sides of the same coin. Firstly, building owners would be well advised to deal with the whole question of collateral warranties at the outset of the project and see that they have bound all the principal parties to the construction process, through their contracts, to give certain collateral warranties when required by the building owner; this should be the case even though the names of the fund/purchaser or tenants are unknown at that stage. Secondly, it may be unwise for architects, engineers and contractors to raise the question of warranties, prior to entering into their principal contracts, with a view to avoiding any binding obligation to enter into warranties.

Contractual Obligation to give Warranties

9.3 The type of clause that could be incorporated into the principal contract to create this obligation is discussed at 6.31 to 6.36. Where such clauses are incorporated by the building owner into tender documents or proposed contract documents, then the tenderer would be well advised to check the provision particularly carefully, including the warranty itself. In contractor's organisations, the estimating department may not be best placed to pick up the risks of collateral warranties and/or their wording and the dangerous consequences that can flow from badly drafted warranties or warranties that have been well drafted but with onerous conditions. Contractor's estimators would be well advised to have such provisions carefully vetted by in-house or external lawyers.

9.4 Where a binding contractual obligation has been entered into to give warranties, then it is likely that, on the failure of that party to give the warranty, the other party to the principal contract could apply to the High Court for an order of specific performance requiring the warranty to be entered into. For such an application to be successful, the wording of the contractual obligation will have to be clear and unambiguous, and the terms of the warranty will have to be certain. A provision that simply purports to require a warranty to be given, but on terms that are not described, will be insufficient to enable an application for specific performance to be successfully made to the court.

Commercial Pressure

9.5 In most cases, it is the commercial pressure to sign warranties that is one of the dominant factors in the decision as to whether or not they should be given. In other words, if warranties are not agreed to be given by a party, then that party may not obtain the project. At a much lower level, there may be such a good working relationship between the parties, with on-going work, that the pressure to give warranties is entirely reasonable. However, in the absence of a binding obligation, the party who is asked to sign warranties is in a much better negotiating position on the terms of the warranty and, perhaps, the extent of the people to whom warranties are to be given. There is little point in the party giving a warranty permitting commercial pressure to force him in to a position where he enters into over-onerous obligations, which are not collateral to the principal contract, or obligations that put at risk his professional indemnity insurance cover or the future financial viability of his business.

COMMERCIAL BALANCE

9.6   The reality is that the purpose of a contract is to allocate risk between the parties; in theory, that allocation of risk is a matter of negotiation between the parties Usually, in collateral warranties, there is not a balancing of risks — most of the risks are undertaken by the person giving the warranty and the negotiation is about the nature and extent of the risks that that person is prepared to undertake. See also 9.7 below.

LEGAL COSTS AND CONSIDERATION

9.7   On lengthy non-standard collateral warranties, the legal costs involved in the original drafting and subsequent vetting, negotiation, re-drafting and producing the final version can be extensive. Indeed, on smaller projects, or sub-contracts on larger projects, the legal costs can be out of all proportion to the value, in terms of profit, of the principal contract. There is no reason in law why the collateral warranty obligations should be given by a professional person or a contractor free of actual consideration in money. Although it is unusual to agree such a 'signing fee', the reality is that the developer, fund and tenant are gaining real benefits through a warranty. The commercial reality suggests a fairly substantial fee might be appropriate. Regard might be had to the financial benefit to the other party; what the premium might be on 10 years' non-cancellable building insurance (which might not be needed if there are collateral warranties); what the market will bear; the future liability risks and any increased professional indemnity insurance premiums, now and in the future by reason of collateral warranties. Indeed, the Royal Incorporation of Architects in Scotland (RIAS) suggest to their members that a charge of 10% of the architect's fees on the project is reaonable, and, that anything less than £1000 would not be reasonable. It goes without saying that many developers, funds and tenants would disagree with the RIAS suggestions as to levels of fee; for example, an architect on a 4% fee on a £10m project would receive a fee, on the RIAS basis of £40,000 in respect of a warranty.

If there is a contractual obligation to enter into a warranty, and there **9.8** has been no prior agreement as to legal costs or some provisions set out in the obligation as to legal cost, then each party will have to bear its own legal costs. However, if there is no legal obligation to enter into a warranty, the person asked to give the warranty is in a rather better position. If the commercial situation enables him to do so, he can say to the other party: 'I am not obliged to give you a warranty but if you wish me to consider giving one, then I must ask you to pay my reasonable legal costs that I will incur in dealing with it and a signing fee of £. . . .' The other party, particularly where it is a developer with prospective tenants who will take a lease if they are given warranties, is likely to be susceptible to that kind of suggestion. Such an agreement as to legal cost can speed the process to agreement of the warranty dramatically — it is simply the case that when one party is picking up the legal costs of two parties, he is likely to be keen to see resolution with the minimum of time, and therefore expenditure of costs.

NEGOTIATING AND INSURANCE

Where the party asked to give a warranty in relation to design has **9.9** professional indemnity insurance, then particular regard should be had to the matters set out in Chapter 7. Some further points bear making in relation to negotiating warranties where there is a professional indemnity insurance aspect.

Where there is an endorsement to the professional indemnity policy in **9.10** the form of, or similar form to, that set out at 7.34 then the person giving the warranty must keep the scope of the warranty within the parameters set by the endorsement to his policy. Indeed, the fact that those parameters are so set by the policy is a very useful tool in negotiations by that party with the person seeking the warranty. He should argue that there is little point in him entering into a collateral warranty that would take him outside the cover of his professional indemnity insurance policy — it is in the interests of all parties to warranties that insurance cover is available in the event of a claim. It is no answer for a developer, fund, purchaser or tenant to say that insurance is a matter wholly for the person giving the warranty; whilst that is true as a matter of law, it flies in the face of the commercial reality in the event of a claim.

Where the insurance policy has no endorsement in relation to collateral **9.11** warranties, or where a warranty cannot be agreed within the terms of the endorsement, then it is absolutely essential to refer all such warranties to insurers for approval or otherwise prior to them being entered into. In order to minimise the workload on insurers, it is sensible to get a draft into its final form before putting it to insurers, rather than constantly referring all the interim drafts; insurers, when faced with such non-standard warranties, will either refuse cover, agree to cover without any extra premium, or agree to cover with extra premium payment. The insured should clarify with the insurer whether that extra premium payment is a one-off in that year of insurance or whether the additional premium is likely to reflect in premiums for subsequent years of insurance. In any event, it is necessary to disclose warranties that fall outside any agreed endorsement to the policy to insurers at the time that insurance is being taken out and at any renewal whether with the same insurer or a different insurer (see 7.8 to 7.11).

WARRANTIES MUST BE SIGNED

9.12 Unilateral gratuitous promises, save perhaps where given in a deed, are not binding in English law. As a matter of practicality, for there to be certainty in a warranty, it must be signed and/or sealed by all the parties to it. Where there is a novation agreement, with three parties, then all three parties must sign and/or execute the warranty as a deed.

9.13 As to the new statutory provisions in relation to the execution of deeds, see 1.46 to 1.48.

THE GIVERS, RECEIVERS AND CONTENTS OF WARRANTIES

9.14 In order to ascertain what warranties are required from whom and what terms they should contain, it is necessary to review the intentions of the developer and the method of procurement of the project.

The Intentions of the Developer

9.15 In setting up the collateral warranty arrangement for a project, it is essential to look first at what the developer's intentions are. Is there to be a fund providing finance for the project and what are their requirements as to collateral warranties, if any? Is there to be a provision for the possibility that the development might be sold during construction or on completion? If it is to be sold during construction, then consideration should be given to inserting an obligation in all the principal contracts (e.g. architect, engineer, quantity surveyor and contractor) requiring that party to enter into a novation agreement, in the form of a draft that should be attached, in the event that the building owner so requires; this is in addition to considering warranties. Is the building to be occupied by the building owner or let by him on full repairing leases to tenants or sold to a purchaser on completion for his occupation or for letting to tenants of the purchaser? There is little point considering who should be required to give warranties, on what terms, and to whom without looking first at the way the project is to be dealt with.

Once the developer's intentions are established, it is a relatively simple **9.16** task to prepare a list of the principal contract and collateral warranty arrangements that will be required. Sometimes this can be best explained by preparing a chart of the contractual arrangements in diagrammatic form (such as Figure 1 and Figure 2 in Chapter 1). Before this can be done, however, it is necessary to look at the method of procurement of the project.

Method of Procurement

The method of procurement will affect what warranties are required from **9.17** whom and the contents of those warranties.

JCT 80 or IFC 84 or Similar
On these types of projects, the building owner will engage the architect **9.18** and the engineer to provide the design. A main contractor will be engaged by the building owner to carry out the construction of the design provided by the architect, and through him, the engineer. There may be, but are not always, nominated sub-contractors who will carry out design for the building owner (usually on the terms of the standard form NSC/2) and who will carry out the construction of their work under a sub-contract with the main contractor (usually in the form of NSC/4).

For the purposes of collateral warranties, similar issues arise in relation **9.19** to management contracts.

JCT 'With Contractor's Design' 1981
Under this form of contract, the building owner sets out his requirements **9.20** in a document which is called in the contract 'the Employer's Requirements'; the design and build contractor puts forward his proposals in a document called 'the Contractor's Proposals'. Both these documents taken together form the description of the work to be carried out for the

purposes of the contract. It follows that, on the face of it, the contractor is undertaking all the design and all the construction work and only he should be required to give warranties. In practice, the contractor will usually sub-let the design to architects and engineers and in some cases (such as mechanical and electrical services) to sub-contractors. In theory there will be a chain of contracts for liability through collateral warranties given by the contractor so that, for example, the tenant can sue the contractor who in turn can join in the architect or engineer or sub-contractor responsible. However, if the contractor goes into liquidation, there will be a break in the chain of contracts and the tenant will be unable to proceed against the architect, the engineer or the sub-contractors in tort. It is for this reason that warranties are often required on design and build projects from the architect and engineer of the contractor, as well as some of the sub-contractors. Clearly a decision needs to be made as to which of the sub-contractors — all of them might be inappropriate.

Construction Management

9.21   There are no standard forms for construction management projects. Usually, the building owner will engage the architect, the engineer and the construction manager; each trade contractor will then enter into a contract with the building owner but on the basis that the construction manager has the right to supervise and direct the trade contractors. These projects are a potential nightmare from a collateral warranty point of view. Clearly the architect and the engineer can be asked to give warranties in respect of design and other matters without too much difficulty. As to the trade contractors, it may be inappropriate to ask for warranties from all of them but clearly the major trade contractors will be asked to provide warranties. As to whether or not a warranty should be taken from the construction manager, this will depend on the duties that he has under his construction management contract. If those duties include, for example, inspecting the work of trade contractors to ascertain that they are carrying out work in accordance with the trade contract then collateral warranties may well be appropriate.

Each construction management contract should be looked at individually to see whether the duties undertaken are duties that are appropriate to the giving of collateral warranties. It may be preferable for landlords and tenants to deal with these kinds of projects by special provisions in the lease (see 10.2 to 10.9) rather than by trying to secure large numbers of collateral warranties, particularly where the completed development is multi-tenanted, like a shopping centre — 30 trade contractors and 20 shop units would lead to 600 warranties (leaving out of account the architect, engineer and construction manager).

Which Terms?

The following tables are designed to assist (but not be definitive) in deciding **9.22**
what terms should go into which warranties on the basis of the different
methods of procurement. 'Purchaser' in these tables is used to mean a
purchaser who has agreed prior to completion of the project to purchase,
completion of the sale to take place *on completion* of the project. Consideration should be given to the general points on drafting at 8.2 to 8.17.

JCT 80/IFC 84 **9.23**

TYPE OF TERM	PARAGRAPH REFERENCE	JCT 80/IFC 84			
		Architect Engineer	QS	Contractor	Major sub-contractors
Skill and care in all duties	8.19	FP?	FP?	—	—
Design	8.19–8.29	FPT	—	—	FPT?
Workmanship	8.32–8.33	—	—	FPT	FPT
Deleterious materials	8.34–8.40	FPT	—	FPT	FPT
Copyright	8.45–8.48	FP(?T)	FP	—	FPT?
Insurance	8.49–8.54	FPT	FP	—	FPT?
Novation	8.55–8.61	EFP	FP	EFP	—
Assignment	8.62–8.67	FPT	FP	FPT	FPT
Other parties	8.68–8.71	FPT?	FP?	FPT?	FPT?
Limiting liability	8.72–8.77	FPT?	FP?	FPT?	FPT?
Limitation of action/indemnity	8.78–8.80	FPT?	FP?	FPT?	FPT?
Delay in design	8.81–8.83	E	—	—	E
Dispute resolution	8.84–8.87	FPT?	FP?	FPT?	FPT?
Notice provision	8.88	FPT	FP	FPT	FPT
Other matters	—	FPTE?	FP?	FPTE?	FPTE?
Under hand/under seal	8.89	?	?	?	?

Key: F = Fund, P = Purchaser, T = Tenant, E = Building Owner
   ? = To be considered in each case

9.24 *JCT 'With Contractor's Design'*

TYPE OF TERM	PARAGRAPH REFERENCE	JCT 'WITH CONTRACTOR'S DESIGN'			
		Contractor	Major sub-contractors	Architect Engineer	*Employer's agent
Skill and care in all duties	8.19	F?	—	—	FP?
Design	8.19–8.29	FPT	FPT	FPT?	—
Workmanship	8.32–8.33	FPT	FPT	—	—
Deleterious materials	8.34–8.40	FPT	FPT	FPT	FP?
Copyright	8.45–8.48	FP(?T)	FP(?T)	FPT?	FP?
Insurance	8.49–8.54	FP(?T)	FP(?T)	FPT?	FP?
Novation	8.55–8.61	EFP	—	—	EFP
Assignment	8.62–8.67	FPT	FPT	FPT	FP
Other parties	8.68–8.71	FPT?	FPT?	FPT?	FP?
Limiting liability	8.72–8.77	FPT?	FPT?	FPT?	FP?
Limitation of action/indemnity	8.78–8.80	FPT?	FPT?	FPT?	FP?
Delay in design	8.81–8.83	—	—	—	—
Dispute resolution	8.84–8.87	FPT?	FPT?	FPT?	FP?
Notice provision	8.88	FPT	FPT	FPT	FP
Other matters	—	FPTE?	FPTE?	FPTE?	FP?
Under hand/under seal	8.89	?	?	?	?

Key: F = Fund, P = Purchaser, T = Tenant, E = Building Owner
? = To be considered in each case
* Whether or not any warranty will be required from the Employer's Agent will depend on the scope of his delegated authority from the Employer (see Article 3 of the JCT 'With Contractor's Design' Form)

Construction Management 9.25

TYPE OF TERM	PARAGRAPH REFERENCE	CONSTRUCTION MANAGEMENT			
		Architect Engineer	QS	Main trade contractors	Construction manager
Skill and care in all duties	8.19	FP?	FP?	—	FP?
Design	8.19–8.29	FPT	—	FPT?	FPT?
Workmanship	8.32–8.33	—	—	FPT?	FPT?
Deleterious materials	8.34–8.40	FPT	—	FPT	FPT?
Copyright	8.45–8.48	FP(?T)	FP	FPT?	FPT?
Insurance	8.49–8.54	FPT	FP	FPT?	FPT?
Novation	8.55–8.61	EFP	FP	EFP	EFP
Assignment	8.62–8.67	FPT	FP	FPT	FPT
Other parties	8.68–8.71	FPT?	FP?	FPT?	FPT?
Limiting liability	8.72–8.77	FPT?	FP?	FPT?	FPT?
Limitation of action/indemnity	8.78–8.80	FPT?	FP?	FPT?	FPT?
Delay in design	8.81–8.83	E	—	E	E?
Dispute resolution	8.84–8.87	FPT?	FP?	FPT?	FPT?
Notice provision	8.88	FPT	FP	FPT	FPT
Other matters	—	FPTE?	FP?	FPTE?	FPTE?
Under hand/under seal	8.89	?	?	?	?

Key: F = Fund, P = Purchaser, T = Tenant, E = Building Owner
      ? = To be considered in each case

BPF FORM OF AGREEMENT FOR COLLATERAL WARRANTY

9.26 Several of the construction professional bodies have from time to time produced warranties; an example was the RIBA form of collateral warranty to be given to a fund. However, the RIBA now regard that agreement as being superseded by the BPF Form of Agreement for Collateral Warranty for use where a warranty is to be given to a company providing finance for a proposed development, known by the acronym CoWa/F. This form is designed for use by architects, engineers and quantity surveyors who are asked to give warranties to a fund. It has been prepared and approved for use by the British Property Federation, the Association of Consulting Engineers, the Royal Institute of British Architects and the Royal Institution of Chartered Surveyors. This warranty is published by the British Property Federation Limited and the copyright is owned jointly by the BPF, the ACE, the RIBA and the RICS with whose kind permission the warranty together with the 'General advice' and their 'Commentary on clauses' is reproduced in Appendix 1 to this book.

9.27 The BPF warranty is not, of course, appropriate for tenants or purchasers but the intention of the parties who prepared the warranty is to move forward to try to agree warranties for other uses. At the same time, contractors and sub-contractors have no standard forms of warranty for funds, purchasers and tenants. The Building Employer's Confederation produce no draft forms and it is believed that they have no plans to publish such forms. However, the Joint Contracts Tribunal is considering the production of draft warranties. The Royal Incorporation of Architects in Scotland produce their own form of duty of care agreement which is approved for use in Scotland by Scheme Underwriters to RIAS Insurance Services Limited. This form can be used in Scotland for funds, purchasers or tenants and the RIAS provide some helpful 'Guidance Notes'. A commentary on Scots law contracts is beyond the scope of this book, but the RIAS form is reproduced (with the kind permission of the RIAS, the copyright holder) at Appendix 2 to this book. It must be emphasised that the RIAS form must not be used in England — it is drafted under the law of Scotland. The RIAS form can be used by an architect for a fund, purchaser or tenant.

9.28 It follows from all this that the only form currently available in England and Wales is the BPF Form to be given to funds. The form is available in pads of five from the BPF, the RIBA Bookshops, the Surveyors Bookshop (RICS) and the Association of Consulting Engineers. It was launched in May 1990 after three years of negotiations and at least twelve drafts. The form was referred to the Association of British Insurers and the Committee of London and Scottish Banks prior to publication. At the launch of the form in May 1990, Mr Michael Mallinson, the President of the BPF, said:

'I can see consultants insisting that they will not sign any other form. Since the form has been agreed by the Association of British Insurers and follows discussion with the Committee of London and Scottish Banks, I hope that financial institutions will accept it and not insist on their own forms. I hope that any other forms of collateral warranty proposed by funding institutions can now be buried — perhaps cremated would be a better word.'

It is too early to say whether that wish is being fulfilled in practice.

Commentary on CoWa/F

This form is only appropriate where a warranty is to be given to a company providing finance for a proposed development. **9.29**

The form is only appropriate for members of the professional team, namely, the architect, the engineer and the quantity surveyor, called the 'Firm' in this form. **9.30**

The form is drafted only for use where the architect/engineer/quantity surveyor has been appointed by the developer (called the 'Client' in this form) in circumstances where the client is receiving funding from a funding institution (called the 'Company' in this form). It follows that this form is not appropriate, unless amended, for design and build projects where the professional team has been appointed by the contractor. **9.31**

The form is a tripartite agreement to which each of the firm, the client and the company are parties. The client is a party to the warranty for the purposes of the variation arrangements (which arise for example where the finance agreement between the company and the client has been terminated) and so that the client can undertake that warranty agreements in the same form will be obtained from other parties (see the commentary below on clauses 5, 6 and 12). **9.32**

Recitals

There are three recitals setting out that the company has entered into a 'Finance Agreement' with the client; that the client has appointed the firm under a contract; and that the client has or may enter into a building contract. **9.33**

9.34   There is no particular problem with these recitals save that the form of the building contract that is proposed to be entered into is important because it triggers the definition of 'practical completion' in clause 9. Some building contracts do not use the words 'practical completion' which are very much a creature of the JCT form. An example from another contract is the British Property Federation Edition of the ACA Form of Building Agreement, where the words used are 'Taking-Over'. Parties proposing to use this form of warranty should therefore be aware of the need to know of the form of the proposed building contract, if it has not been entered into at the time the warranty is given, and to make an amendment, if necessary, to the words 'practical completion' in clause 9.

Consideration

9.35   The warranty sensibly provides (immediately above clause 1) for consideration by means of the usual formula, namely, in consideration of the payment of £1 by the company to the firm, receipt of which the firm acknowledges. Although such a provision is probably not necessary where the warranty is to be executed as a deed (but see 9.75), it is most certainly necessary where it is executed under hand. It follows that having this provision in the standard form will avoid the risk of the consideration point being missed where the warranty is to be executed under hand. Incidentally, there is no reason why a substantial fee could not be agreed. The Royal Incorporation of Architects in Scotland suggest 10% of the architect's fee on the project, with a minimum of £1000.

9.36   However, the express words contain no provision in respect of consideration *vis à vis* the client. Consideration is discussed at 1.42 but in essence the usual analysis involves either detriment to the promisee or benefit to the promisor. Indeed, there has to be consideration for each promise not for the contract as a whole. In relation to the client, therefore, it is necessary, in the absence of an express statement as to consideration, to search the warranty to try to find the appropriate detriment and/or benefit. In so doing, the propositions as to consideration where there are three parties to an agreement set out at 1.44 need to be borne in mind. It is unfortunate that it should be necessary to carry out this analytical exercise, which in the event produces an uncertain result: it is hard to find the appropriate detriment and/or benefit for the client's promises *vis à vis* the firm. There is therefore doubt whether in that respect there is consideration in this warranty. This issue could be put beyond doubt by amending the consideration provision to say that each party pays every other party £1, receipt of which is acknowledged, and in consideration of the mutual undertakings set out in the warranty.

These issues are further considered at 9.52, 9.74 and 9.75.   **9.37**

Clause 1

The purpose of this clause is to fix the firm with a duty in contract to the   **9.38**
company; there are alternative provisions in this clause appearing in
square brackets. This is to enable the warranty to be amended to
correspond with the contract to which this warranty is collateral. For
example, where the architect is engaged on Architect's Appointment, the
words '[care and diligence]' should be deleted in the warranty. Where the
engineer is appointed on the basis of the ACE Conditions, the words '[and
care]' should be deleted. The RICS Standard Form of Agreement for the
appointment of a quantity surveyor does not contain a provision as to the
standard of the duty to be exercised. It will be that of reasonable skill and
care by implication of a term to that effect. Therefore, where this form is
used for a quantity surveyor, clause 1 of the warranty should be amended
by deleting the words '[care and diligence]'. If some other conditions of
engagement are used, then clause 1 should provide the same words by
way of duty as appear in those conditions. The BPF publish their own
Conditions of Engagement for Consultant's Work which are derived
from the ACE Conditions of Engagement, Agreement 3. The BPF
Conditions of Engagement have been approved by the ACE and
published after consultation with the RIBA and they provide at clause
5.1:

'The Consultant shall exercise all the skill, care and diligence in the
discharge of the Services to be expected of an appropriately qualified
and competent consultant experienced in carrying out services for a
project of a similar scope, purpose and size to the Project and having
regard to the dates and periods stated in the Priced Programme.'

Although based on the ACE Conditions, the words after 'to be expected   **9.39**
of . . .' are additional to the words in the ACE Form. Arguably, these
additional words impose a higher duty than the reasonable skill and care
to be expected of an ordinary competent firm exercising their profession.
It might be said that the BPF collateral warranty is not consistent with the
BPF Conditions of Engagement in this respect and consideration might
be given to amending clause 1 for consistency where the BPF Conditions
of Engagement are used.

Clause 1 has a proviso that the firm '. . . shall have no greater liability   **9.40**
to the Company' than it would have had if the company had been named
as a joint client under the conditions of engagement of the firm by the

client. The words 'no greater liability' may give rise to some difficulty of
proper legal interpretation (see 8.76). Usually, the issue is whether there
is liability or not and, if there is liability, what is the *quantum* of damage
that flows from that liability. The wording of this warranty gives rise to
some uncertainty as to its proper legal construction. It may create some
limitation on liability but may or may not be successful in limiting the
consequences of that liability in terms of recoverable damages. Indeed,
the Court of Appeal in two recent cases has drawn a clear distinction
between liability and damage in relation to the interpretation of express
provisions in contracts: *Beaufort House Devlopments Limited* v. *Zimmcor
(International) Limited* and *Rosehaugh Stanhope Broadgate Phase 6 plc* v.
Redpath Dorman Long Limited.

9.41   This clause provides a very wide ranging duty to the company, namely,
reasonable skill and care in the performance of *all* the firm's duties under
their conditions of engagement. Taking the Architect's Appointment as
an example, those duties include:

- each of the services to be provided (which could include inception,
  feasibility, outline proposals, scheme design, detailed design,
  production information, bills of quantities, tender action, project
  planning, operations on site and completion, being Work Stages A
  to L)
- not making any material alteration to the approved design without
  consent
- informing the client that the expenditure or the building contract
  period are likely to be materially varied
- co-ordination of the design
- inspecting progress and quality of the building works, and
- the exercise of suspension and termination rights (see clauses 3.3,
  3.4, 3.7, 3.8, 3.10, 3.20, 3.22 and 3.23).

It should be noted that the architect's duties in relation to Work Stages K
and L include administering the building contract — that of itself includes
various duties such as certification of sums due to be paid to the
contractor, the certification of loss and expense (even though the
ascertainment may have been delegated to the quantity surveyor),
granting extensions of time, certifying practical completion and giving
non-completion certificates. In respect of all these functions, therefore, an
architect will, on this form of warranty, have entered into a contractual
obligation with the fund to perform all those duties with reasonable skill

and care. This warranty is not limited to the exercise of reasonable skill and care in the design.

Clause 2

This is the deleterious materials provision which is intended to be deleted only where the warranty is given by a quantity surveyor.   **9.42**

The obligation is not to specify certain stated materials (at (a) to (e) in the clause) but it is subject to a reasonable skill and care test. The warranty does not purport to require the firm to see that such materials are not incorporated into the works. There is a savings provision so that this duty will not apply where the client has authorised the use of those materials (either in writing or, if oral, confirmed in writing). Clause 2 begins with the words 'without prejudice to the generality of clause 1' and the intention of these words is to provide that the detailed deleterious materials provisions in clause 2 do not cut down the overriding general duty, set out in clause 1, to exercise reasonable skill and care in the performance of the duties. It follows that an express approval by the client to the use of any of the specified materials does not bind the company if the company can establish a breach of clause 1 by the firm.   **9.43**

In the form at clause 2(f), a space is left for completion by the parties, if they wish, and the marginal note reads 'further specific materials may be added by agreement'. It is likely that this space will be used, not only for further specific materials, but also for a sweep up clause of the type discussed at 8.39 to 8.40 and all the issues discussed there should be considered if a sweep up clause is inserted in this form of warranty.   **9.44**

Clause 3

This provides that the company cannot give any instructions to the firm except where the company is stepping into the shoes of the client. As to that exception, see the commentary on clauses 5 and 6 at 9.48 to 9.58 below.   **9.45**

Clause 4

Here, the firm acknowledges that it has been paid all the fees and expenses due and owing to it by the client at the date of the warranty agreement. That part of this clause probably has two effects: firstly, it gives a firm that has not been paid what it is owed at the time it is asked to sign the warranty agreement the opportunity to refuse to sign unless and until it has been paid. Secondly, if the firm were to sign in circumstances where they had not been paid up to date, then the firm could not recover their outstanding fees from the company under the provisions of clause 7,   **9.46** following the company taking over from the client under either of clauses 5 or 6. It follows that every firm should be particularly careful to check that they have been paid up to date prior to signing this form of warranty. No provision is made for what is to happen if there is a *dispute* about fees at the time the firm are asked to sign the warranty.

9.47   This clause further provides that the company has no liability to the firm in respect of fees and expenses unless and until they have taken over from the client under clauses 5 or 6.

Clauses 5, 6 and 7

9.48   These provisions set out the circumstances in which the company are entitled to take the place of the client under the contract between the firm and the client and what the consequences of such action are to be. It should be noted that the company is given these rights either for itself or its 'appointee'. The effect of this provision is intended to be that the company can require the firm to accept a third party of the company's choice, but see below at 9.54 to 9.56. In practice, where the company is a bank or pension fund it is highly unlikely they would wish to be involved in the detailed work that is necessary to be the client for the purposes of conditions of engagement. In such circumstances, a bank/pension fund are likely to wish to appoint a project manager or similar to perform these functions. The firm is given no right of reasonable objection to the appointee.

9.49   There are two circumstances in which the company can 'require . . . [the Firm to] . . . accept the instructions of the Company or its appointee':

(1) Where the finance agreement (between the company and the client) has been terminated (clause 5). The written notice of the company is, for the purposes of this clause, to be conclusive evidence that the finance agreement has been so terminated.

(2) Where the firm has a right to terminate or treat as having been repudiated the contract with the client (clause 6). The procedure is that where such a right arises on the part of the firm, they agree in the warranty that they will not exercise that right unless they have first given at least 21 days' notice in writing to the company. The company may during that 21 days give notice that either they or their appointee are going to give instructions in the future. If the company do give that notice, then the firm's rights to terminate the contract with the client or treat it as having been repudiated cease.

9.50 Whether action is taken by the company under clause 5 or clause 6 is entirely at their discretion; it follows that the firm cannot force the company to take any action under either of these clauses and that clauses 5 and 6 are a benefit for the company.

However, if the company does take action under clauses 5 or 6, then **9.51** clause 7 applies. Clause 7 provides for several matters:

(1) The company or its appointee become liable for payment of the firm's fees and that includes any fees outstanding at the date of the notice, but see the commentary on clause 4 at 9.46 above.

(2) The company or its appointee take on the performance of the client's obligations under the contract between the client and the firm (but on the words there has to be doubt as to whether the company takes over the *existing* obligations of the client, other than fees, or whether the effect of this provision is simply to deal with future obligations. This is an important point as to which there should be no doubt in the drafting).

(3) The contract between the client and the firm continues in full force and effect.

(4) The firm becomes liable to the company or its appointee under the contract between the client and the firm in place of the firm's liability to the client.

(5) Where the company appoints an appointee in the notice, then notwithstanding the fact that the appointee is primarily liable for the payment of the firm's fees, the company is liable to the firm 'as guarantor'.

The provisions in clauses 5, 6 and 7 are a variation of the contract, not a **9.52** novation. Variations (as novations) require consideration, unless, perhaps, the document is executed as a deed. This variation needs consideration and there is doubt as to whether there is consideration *vis à vis* the client (see 9.36).

All the provisions in clause 7 are preceded by the words: 'It shall be a **9.53** condition of any notice given by the Company under Clauses 5 or 6. . . .' These words are capable of two interpretations given the proposition that the precise requirements as to the contents of a notice are not set out. The first is that if a notice is given, then the matters set out in clause 7 apply; the second is that a notice is not a valid notice under the warranty unless it recites on its face all the matters set out in clause 7. The latter is probably the better view. If that is right, it follows that a company giving a notice under clause 5 or 6 should recite the provisions of clause 7 in the notice in order to put this point beyond doubt. The procedure for giving notices is in clause 13.

9.54   However other difficulties arise on the wording of the provisions of clauses 5, 6 and 7 in relation to the company appointing an 'appointee'. It is not possible for a party to be a party to a contract in the absence of agreement between the parties. The appointee is not a party to the warranty at the outset, unlike the company. The appointee is not the agent in law of the company; the language of clause 7 is inconsistent with such a view in that the appointee takes on the obligations of a principal, namely the obligations of the client to the firm and the obligation to pay the firm; indeed the company are said to be a guarantor of payment of the fees to the firm by the appointee. Against this background, it is submitted that the appointee provisions of this warranty agreement do not work in law to create the intended rights and duties between the appointee and the firm.

9.55   An incidental difficulty arises where the warranty and/or the appointment of a firm have been executed as a deed by the original parties. Such a deed cannot be varied unless the variation itself is contained in a further deed. On any view, the appointment of an appointee is a variation, and, the notice under clauses 5 and 6 is not a deed. A further difficulty is that, in any event on the provisions of clauses 5, 6 and 7, without more, the appointee cannot be bound by way of deed. One way to avoid all these potential difficulties in relation to an appointee is to amend the warranty so that where the company wish to appoint an appointee, the firm have an obligation on receipt of notice from the company, to enter into a variation agreement in the form of a draft attached to the warranty agreement — the form of that draft agreement could be easily derived from clauses 5, 6 and 7 and the attestation clause in the draft could make it clear whether the variation agreement was to be executed under hand or by deed. The fact that the identity of the appointee is not known at the date of the collateral warranty agreement would not prevent this procedure from being effective.

9.56   It follows that firms in receipt of notices under the provisions of clauses 5, 6 and 7 of this warranty should give them very careful consideration on all the points mentioned above and in particular, where there is an appointee, whether or not that appointee is bound in law to the obligations that are set out in clause 7 and whether or not the firm will be in breach of its contract with the client (see 9.74 below) if it acts on the notice.

9.57   The language of these clauses is that of 'obligations', 'instructions' and 'liability'. In novation agreements, the usual language seeks to deal with the benefits and burdens of each of the three parties. Looking at the provisions of these clauses from the standpoint of benefits and burdens, and the express words in clause 7, it is the case that the client does not obtain a discharge in respect of the burdens he has undertaken to the firm. This might be of assistance to a firm seeking to recover fees unpaid at the date of the notice in circumstances where they cannot recover them from the company by reason of the provisions of clause 4 (see 9.46). A further potential problem arises in relation to the provision in clause 7 that the firm and the client remain bound by their contract; in clauses 5 and 6 provision is made for the firm 'to accept the instructions of the Company . . . to the exclusion of the client . . . .' That deals with what the position of the firm is to be, but it does not provide (and neither does clause 7) that the client shall issue no further instructions to the firm. This could lead to difficult issues where the company are disputing the operation of clauses 5 and 6 for *bona fide* reasons. The point could be put beyond doubt by simple amendments.

There is no provision for the client to be served with a copy of any **9.58** notice given under clauses 5 or 6. Given the major changes in the responsibilities of the parties that are triggered by that notice, this is a curious omission.

Clause 8

This clause provides for the company to have a licence to copy and use **9.59** certain documents, the copyright in which remains with the firm. The licence granted is limited to 'any purpose related to the Development' and on the wording of the clause the licence also extends to the use of the documents where there is to be an extension of the development but the rights under the licence do not extend to reproducing the designs contained in the documents for the purposes of building an extension to the development. The clause provides that the firm will not be liable to the company for any use to which the documents are put where that use is for a purpose other than that for which the documents were prepared by the firm; this is clearly an important provision from the firm's point of view. The licence extends to a right on the part of the company to copy the drawings and documents; there is, therefore, no right on the part of the company to require the firm to produce copies at the firm's expense.

These provisions give more rights to the company through the **9.60** warranty than, for example, the architect gives to his client through the Architect's Appointment. Under the Architect's Appointment, the right of the client to reproduce the design by executing the project is subject to conditions which are all cumulative; in particular, the architect has to have reached work stage D or provided detailed design and production information in work stages E, F and G, and, the architect has to have been paid any fees that are due (see clauses 3.15, 3.16 and 3.17 of Architect's Appointment). There are no such restrictions in the licence given to the company under the warranty agreement (c.f. clause 7 of the RIAS form).

9.61   One potentially serious consequence for the firm of the licence contained in this collateral warranty is that if their client becomes insolvent at a time when they have completed their design, but the firm has not been paid, the company can use that design to complete the building without paying the firm's fees because the company has no liability to the firm in respect of fees unless they trigger the provisions of clauses 5 or 6 (which they may be disinclined to do when they will obtain little benefit, having the design already, and may incur substantial liability to the firm for fees).

9.62   This clause also contains reference to an 'appointee'. Indeed, the licence given by this clause is given to the company *and* its appointee. Two points arise. The first is the issue as to whether or not the appointee in clause 8 is the same as the appointee in clauses 5, 6 and 7. As a matter of legal construction it does not follow that it is the same appointee; secondly, if it is not, then this clause will assist the company on the point at 9.61 above.

Clause 9

9.63   For a commentary on this professional indemnity insurance provision see 8.49 to 8.54 above. The obligation referred to in clause 9 is nothing more nor less than an agreement to agree and is unenforceable in law. The same point in relation to the reference to 'appointee' in this clause arises as at paragraph 9.54 above. The 'appointee' in this clause is the appointee under clauses 5 or 6.

9.64   There is a small misprint in the printed form in clause 9 in that no gap has been left in the text for the insertion of the number of years between the words '. . . for a period of' in the third line and '. . . years from the date . . .' in the fourth line. It goes without saying that the period of years must be inserted in this clause as must the amount of the limit of indemnity.

Clause 10

9.65   This clause sets out why the client has agreed to be a party to the collateral warranty — for the purposes of clause 12 (see 9.70 below) and clauses 5 and 6 (see 9.48 above). The substance of this clause is similar to a provision contained in the now superseded RIBA collateral warranty for a fund; the wording of the clause is more akin to a recital than a condition and a question may arise as to whether the clause does as a matter of law give an acknowledgement that the firm will not be in breach of its contract with the client if it acts for the company under clauses 5 and 6, or whether the clause simply states that that is one of the purposes for which the client has agreed to be a party: the purpose not in reality being fulfilled. On the other hand it may be that a court would be disinclined to adopt such a strict legal construction. The point could be put beyond doubt by inserting after the word 'acknowledging', the words 'which the Client hereby does so acknowledge'.

Clause 11

This clause deals with assignment. It is in very different form to the clause that was contained in the now superseded RIBA form of agreement for collateral warranty to a fund. The RIBA form of clause was as follows: **9.66**

'This Agreement may be assigned to one further Company prior to practical completion and within three years hereof, subject to the express consent of the Consultant, which shall not be unreasonably withheld, and subject to the payment of all fees due to the Consultant with credit for payments made. No further assignment will be permitted.'

The move from that fairly restrictive position on assignment to the now apparently much wider provision is one of the reasons that some members of the RIBA are particularly concerned about the BPF warranty (see, for example, the article by Clifford Lansley, Chairman of RIBA Indemnity Research and former Chairman of the RIBA's Liability Committee, *Building Design*, 11 May 1990). **9.67**

Firstly, there can be assignment by the company without the consent of either the client or the firm but notice has to be given by the company to both the client and to the firm (but note there is no procedure in clause 13 for the service of notices on the client). Indeed, the assignment envisaged in the clause is 'by way of absolute legal assignment'. The effect of that provision is that that assignment has to comply with the formalities of section 136 of the Law of Property Act 1925 (see 3.7) and, in particular, there has to be express notice in writing to the debtor (the firm) by the assignor (the company) before a valid legal assignment can come into effect. The assignment envisaged in the clause is an *absolute* legal assignment. This involves an assignment which is not qualified by conditions, is not by way of charge and is in respect of the whole thing in action (see 3.9 to 3.10). Once that absolute legal assignment has been effected, with all its formalities, the company (the assignor) has no right **9.68** to sue on the warranty (see 3.15). There is no prohibition on further assignment by the assignee.

9.69   The clause does not appear to prohibit assignment in some other form; for example, the clause does not say that assignment can only be by the method set out in the clause, or, that there shall be no right to assign unless it is effected in the manner provided in the clause. It does therefore seem likely that there can be an equitable assignment with or without notice to the client or the firm (see the discussion at 3.19).

Clause 12

9.70   This clause provides an undertaking by the client that warranties in the same form 'or in substantially similar form' will be entered into between the company and other parties, which are to be named in the blank space in clause 12. This clause arises for the reasons set out at paragraphs 8.68 to 8.71. Firms will wish to have the contractor as one of the other parties (although the BPF warranty itself will not be appropriate). In view of the potential for reducing the amount of outlay in the event of a claim, it is surprising that the insurers (RIBA, RICSIS and ACE) do not give some *detailed* guidance on this clause in the 'General advice' or 'Commentary on clauses': for example guidance on which other firms and companies should be listed, and having those other warranties by deed if this warranty is by deed.

9.71   The question that arises is what is the effect of this clause in relation to the firm's liability, and if it is liable, the recoverable damages? It is clear that the clause does not create a condition precedent to the firm's liability that those other warranty agreements have been entered into. It is equally clear that this clause does not create any form of net contribution agreement (see 8.71 above).

9.72   In the event that the client does not obtain the warranties that he has undertaken to obtain with the other parties, there will be a breach by him of that obligation. In that event, does the breach give rise to only nominal damages or something more useful to the firm? The issue as to damages will be determined by the rules on remoteness (see 5.7) and whether or not those damages are simply too uncertain to be ascertained. (See 5.27 and 5.28.)

9.73   If the client had entered into the other warranty agreements, then the firm would be entitled to bring contribution proceedings under the Civil Liability (Contribution) Act 1978 against another firm who had given a warranty, provided that other firm is liable in respect of the same damage (section 1(1) of the Act). By reason of the present law of tort, that other firm, if there is no warranty, would not be liable in tort in respect of the same damage. It follows that the failure of the client to obtain the other warranty agreements will have prevented the firm from seeking contribution from the other firm or firms liable in respect of the same damage. Assuming that the firm could satisfy a court that the other firm would have been liable in respect of the same damage had a collateral warranty been entered into, the next issue is whether the amount of that contribution can be ascertained with any certainty by a court. There are some cases referred to at 5.27 and 5.28 which indicate that a court might be prepared to entertain such a relatively speculative calculation. If it did so, then the firm would recover the amount of the contribution, not from the other firm liable in respect of the same damage, but from the client as damages for breach of clause 12. It is to be noted that this will not serve to reduce the damages that the company will recover from the firm.

The issue as to whether there is consideration for the undertaking given by the client in this clause is discussed at 9.36. If there is no consideration, then there will be no obligation on the client as a matter of law to comply with the undertaking set out in clause 12 to obtain warranties from other parties. **9.74**

Finally, it may be possible for a firm to apply to the court for an order for specific performance if the client is in breach of his undertaking under clause 12 to obtain warranty agreements from the other named firms (see 1.49). However, such an application could not be made in circumstances where there is no consideration: that principle of equity applies whether or not the warranty has been executed under hand or as a deed. On an application for specific performance on the client's undertaking in this clause, the words 'or in substantially similar form' may make the prospects of success on the application rather poorer than they might otherwise have been. If, however, a circumstance were to arise where the other firm were content to enter into a warranty agreement, in precise terms, but the client did not (say because he did not want to pay fees or there were a dispute over fees), then the prospects of success would be likely to be much improved. **9.75**

Clause 13

This clause provides for the formalities of the giving of notices but not in respect of the client. It is important to follow the requirements in this clause when giving notices under the warranty. Notices can be given by hand or sent by registered post or recorded delivery. In respect of notices sent by registered post or recorded delivery (but not by hand), the notice is deemed to have been received 48 hours after being posted. The effect of this deeming provision is that a posted notice takes effect even though it **9.76** may not actually have been received by the addressee. There is no provision for service by fax.

9.77   It is surprising that there is no provision for the service of notices on the client in this clause. Whilst this is consistent with clauses 5, 6 and 7 which do not require notices to be served on the client, the operation of clause 11 does require service of notice on the client.

General

9.78   There are alternative attestation provisions for where the warranty is simply to be signed by the parties (under hand) or sealed by the parties. In the event of the former, the limitation period is six years from the date of accrual of the cause of action whereas in relation to the latter it is 12 years from the date of accrual of the cause of action (see Chapter 5 for a discussion). Since the warranty was published, two statutes have changed the law on execution of documents. Where the parties are limited companies executing under their company seals, the printed form is in order. However, where a party is an individual or a partnership executing under seal, or a company that does not have a company seal executing under seal, or a company wishing to execute under hand, amendments will be needed to the attestation provisions (see 1.46 to 1.48).

In addition to those matters, the issue of consideration will arise (see 9.35 and 9.36). In the 'General advice' issued with the form, it is said at paragraph 6 that the warranty should not be signed under seal when it is collateral to an appointment which is under hand. Whilst that approach is sensible for consistency it does not necessarily need to be the case. A company providing finance may well insist on a warranty by deed at a time when the firm have already been appointed under a contract under hand. In this event, careful consideration should be given by the firm and the company as to the effective limitation period that they will obtain — this arises, in particular, in relation to the words in clause 1 of the warranty that the firm shall have no greater liability to the company than it would have had if the company had been named as a joint client under the conditions of engagement (see the commentary at 9.40). It is to be doubted whether those words will constitute an agreement that the warranty by deed will have a six year limitation period. Given the potential difficulties in relation to consideration *vis à vis* the client (see 9.36), the parties may prefer, in any event, to execute the warranty under seal notwithstanding the company may not thereby obtain the benefit of a longer limitation period. Furthermore, there is the issue as to whether specific performance can be obtained (see 9.75).

9.79   At paragraphs 7 and 8 of the 'General advice' issued with the form, some points are made in relation to the professional indemnity aspects of the use of this form. Firstly, there is a reminder at paragraph 7 that professional indemnity policies are annually renewable and that whether or not a claim arising under the warranty will be met by the policy 'will depend upon the terms and conditions of the policy in force at the time when a claim is made'. At paragraphs 8 and 9 it is said that consultants with a *current* policy taken out under the RIBA, RICSIS or ACE schemes will not have a claim refused simply on the basis that it is brought under the terms of a collateral warranty provided it is in the terms of this standard form, unamended. Taking these two propositions in paragraph 7 and 8 together, the RIBA, RICSIS and ACE schemes do not appear to be accepting that a claim made at a date beyond the period of the *current* year of insurance will necessarily be met under the policy in force at that time — that will depend on the terms and conditions then in force. In view of the fact that the 'General advice' does not clearly state what is meant by the words 'consultants with a current policy', every consultant would be well advised to check with his then insurer prior to entering into this warranty whether or not they can do so without express permission from the insurer. Insurers' policy on these issues may change. Consultants who are insured with other insurers should adopt that procedure in any event. See also 9.70 above.

9.80   The 'General advice' and the 'Commentary on clauses' which are printed at the front of the pad of forms with the warranty are set out in Appendix 1 to this book. Those short form notes do not affect the proper legal interpretation of the warranty and are, necessarily, in cryptic form.

Chapter 10

Other Solutions: Present and Future

10.1 The difficulties created by the widespread use of collateral warranties are legion: the length of this book is testimony to that proposition. Those difficulties have prompted many people to ponder whether there could be a better way to regulate these matters, with or without collateral warranties. What follows here is a discussion of some of those possibilities: some of these solutions are available now, and others are contemplating the future.

POSSIBLE SOLUTIONS — THE PRESENT

Commercial Leases

10.2 The terms of commercial leases usually include full repairing covenants on the part of tenants (see 6.2 to 6.9). Where a developer/landlord is willing, and they sometimes though not often are, a great deal could be done to remove the need for collateral warranties by having terms in the lease which put the risk of repair in respect of latent defects on to the landlord, rather than the tenant. Some of these methods are set out below.

Excluding Tenant's Liability to remedy Latent Defects
10.3 This option is that the tenant is subjected to a full repairing obligation in the usual way, but that there is excluded from that obligation any liability on the tenant to remedy latent defects. Such a provision might be:

> 'Provided that nothing in this Lease shall be construed as obliging the Tenant to remedy any Defect of whose existence the Tenant has within the first . . . years of the Term notified the Landlord or any want of repair which is attributable to such Defect and which manifests itself [within such period *or* at any time during the Term].'

Subject to the definition of 'Defect' which is dealt with below, this **10.4** provision would put the liability for the repair of latent defects and any want of repair caused by the latent defect on to the landlord. Tenants looking at such a provision may well wish to exclude from their liability to service charges the cost of remedying such latent defects.

Landlord to remedy Latent Defects

Another option is to have a provision in the lease which puts the **10.5** obligation and the cost of remedying latent defects on to the landlord, rather than the tenant. Such an obligation might be in the following form:

'The Landlord shall at its own expense remedy any Defect of whose existence the Tenant shall within the first . . . years of the Term have notified the Landlord and any want of repair which is attributable to any such Defect and which manifests itself [within such period *or* at any time during the Term].'

Again, subject to the definition of 'Defect', this provision puts the liability **10.6** for the cost of repair in respect of latent defects on to the party, the landlord, who will usually be in the best position in law to make recovery from the parties responsible for the latent defects under the principal contracts between the landlord and the construction professionals/the contractor/sub-contractors.

Payment to Tenant of Cost of Repairs

A further possibility is to provide in the lease that the tenant is under a full **10.7** repairing obligation but that in respect of the cost of repairing latent defects and any consequential repairs, the landlord has an obligation to reimburse the tenant. Such a provision would substantially relieve the tenant's burden, remove the need for collateral warranties, and leave the landlord in a position where he could pursue the parties with whom he is in contract, in respect of the loss that he has incurred under the lease, by reason of the latent defects. Such a clause might be:

'If the Tenant shall in compliance with its obligations under [the repairing covenant] of this Lease carry out any works to remedy any Defect of whose existence the Tenant shall within the first . . . years of the Term have notified the Landlord or to remedy any want of repair which is attributable to any such Defect and which manifests itself [within such period *or* at any time during the Term] then provided that the Tenant prior to carrying out such works shall obtain the approval of the Landlord to the nature and extent of the works [such approval not to be unreasonably withheld [or delayed]] and complete such works to the [reasonable] satisfaction of the Landlord the Landlord shall pay or repay to the Tenant the costs and expenses of and incidental to such works following production to the Landlord of all relevant receipts and invoices or other reasonable evidence of such costs and expenses.'

Definition of 'Defect'

10.8   In all the above provisions, to make the clauses workable, there has to be a careful and clear definition of 'Defect'. It will be necessary to provide for the definition to include negligent design, workmanship, materials and plant not in accordance with the building contract, and negligent supervision of the construction of the building. Clearly, the tenant will wish to have as wide definitions as possible in those respects. On the other hand, the landlord is likely to want to exclude any defect which was patent, as opposed to latent, at the date of the lease. This could be achieved by excluding from the definition of 'Defect' any defect which was visible or ought reasonably to have been visible to a competent surveyor at a time immediately before the date of the lease. A precedent for this kind of definition can be found in the *Encyclopaedia of Forms and Precedents: Volume 22: Landlord and Tenant (Business Tenancies)*, Form 61.

Supplemental Deed

10.9   By means of a deed supplemental to the lease, the landlord can be obliged to take all reasonable steps, including proceedings, to enforce against third parties (e.g. the architect, the engineer and/or the contractor) any rights which the landlord has in respect of the latent defects. In such an agreement, it is necessary to deal with the proceeds of such litigation (i.e. are they monies held in trust for the tenant by the landlord?) the definition of defects and who is to pay the costs of any proceedings that may be necessary. In order that a tenant could feel reasonably secure in entering into such an arrangement, it is important that his legal advisers should investigate before entering into the lease what rights the landlord has against third parties. This may involve inspection of the building contract and the conditions of engagement of the professional team, as well as, in appropriate cases, collateral warranty agreements between designing sub-contractors and the landlord, and any other rights the landlord may have in sub-contracts. A draft Supplemental Deed of this type can be found in the *Encyclopaedia of Forms and Precedents: Volume 22: Landlord and Tenant (Business Tenancies)*, Form 21.

Purchasers

The problems of *caveat emptor*, 'buyer beware', are discussed at 6.19 to 6.22; collateral warranties can often be avoided *vis à vis* a purchaser, and the method is determined by whether or not the purchase is completed before or after all the obligations (save in respect of latent defects) have been fulfilled under the principal contracts with the building contractor and the professional team. **10.10**

If the purchase is to be completed before completion, then the appropriate way to give the purchaser the rights that he needs is for there to be a novation agreement in respect of each and every principal contract (see 8.56). Draughtsmen in preparing such agreements should have particular regard to the state of outstanding fees *vis à vis* the professional team and, in relation to the contractor, interim certificates/uncertified work in progress and retention monies. Where the purchase is completed at a later date (say after the release of retention to the main contractor and the payment of all fees due to the professional team) then the purchaser can be best protected by an absolute legal assignment of the benefits of the developer under each and every principal contract. **10.11**

In order that the developer and purchaser can have the right to proceed by either novation or assignment, the terms of the principal contract may require amendment. For example, where there is to be a novation, it would be appropriate for the principal contract to contain a provision requiring the other party at the behest of the developer to enter into a novation agreement in the form of a draft attached. Where there is to be an assignment, the principal contract should provide that that assignment can be effected by the developer without the consent of the other party to the principal contracts, or with their consent, which consent is not to be unreasonably withheld. **10.12**

However, where the purchase is agreed during construction but completion of the sale is to take place on practical completion of the building works, collateral warranties may well be necessary (see 6.19 to 6.25 and 9.22 to 9.25). **10.13**

Insurance

The insurance industry has responded to some of the difficulties set out in this book. In particular, ten year non-cancellable insurance on the **10.14** building (subject to various restrictions and extensions) is now more widely available than it has been. The NEDO report 'Building Users' Insurance Against Latent Defects' has stimulated discussion and action in this area. NEDO is the independent office which supports and assists the National Economic Development Council and its sector groups and working parties. It plays a key role in promoting change.

BUILD

10.15   In 1988, a report was produced by the Construction Industry Sector Group of the National Economic Development Council. The report is called 'Building Users' Insurance Against Latent Defects' known by the acronym BUILD. The committee that produced the report consisted of many distinguished people, (including architects, engineers, contractors, property developers, insurers and representatives of government), under the chairmanship of Professor Donald Bishop. The report contains an excellent analysis of the risks of building development to the parties involved in it. Amongst the many points that arise from this report is the recommendation that there should be 'BUILD insurance'. The essence of such insurance is that it should provide protection for a period of ten years from the date of practical completion, on the basis that the insurance is a non-cancellable material damage policy against specified latent defects and damage. It is suggested that the cover initially would be limited to the structure, including foundations, the weathershield envelope and, optionally, loss of rent.

10.16   The policy would be taken out by the developer at a very early stage, at least before any work is started on the site. The policy would be transferrable to successive owners and to tenants of the whole building; in a situation where the whole building is not let to one tenant, the report envisages that such tenants would be indemnified by the landlord on a back-to-back basis with the terms of the BUILD policy held by the landlord. There would be a single premium to cover insurance and the necessary risk assessment and verification of the design and construction by independent consultants appointed by the insurer. There should be provision for inflation in building costs in respect of the cover, and realistic deductibles (excess). A further recommendation of the report was that insurers should waive their rights of subrogation without which insurers would be in a position to seek to recover their outlay from the party liable to the developer: for example, the architect, the engineer and/or the contractor. At the time of publication of the report, this aspect was of considerable concern to insurers because it would create, in effect, a no fault insurance scheme (but see below at 10.29).

At the time of the report, there were three underwriters in the market: **10.17**
Allianz, Norman and SCOR. There are now more underwriters in the
market and the Sun Alliance have recently introduced a policy along the
lines envisaged by BUILD (see 10.18 to 10.25). The indication in BUILD of
the premium, including the verification costs, was that they would be of
the order of 1.3% to 1.7% of the rebuilding cost (although *currently*
premium levels can be lower). Although a small percentage, the cost in
money terms can look substantial: for example, on a £10 million
rebuilding cost, the premium and other associated costs would be of the
order of £150,000 if the rate were 1.5%. Some developers consider that
level of premium a high cost to pay when they are in a position to obtain
tenants who are prepared to take the risk on full repairing covenants, with
collateral warranties, all free of cost to the developer. On the other hand,
if the developer comes to sell the building, the fact that there is such
insurance may enable a better price to be achieved, or alternatively, a sale
to be achieved which might not be achieved in the absence of the cover.
Furthermore, in times of over-supply of commercial buildings in the
market place, it is conceivable that tenants may be more attracted to
buildings with such cover than buildings without the cover. Developers
might be rather more keen on BUILD type cover if they were regularly
asked to pay substantial fees to the professional team, the contractor and
sub-contractors as consideration for the signing of warranties (see 9.7).

Building Defects Insurance
There are now several insurers writing building defects type insurance; it **10.18**
simply is not possible to examine all the policy options available from all
the insurers and underwriters in this field of insurance. However, it may
be helpful to look in a little detail at one such policy: Sun Alliance
Building Defects Insurance. This policy has been available since late 1989.

The cover under this policy enables the insured to recover the cost of **10.19**
repairing damage, the cost of remedial action to prevent imminent
damage, professional fees, the costs of debris removal and site clearance,
and the extra costs of reinstatement to comply with public authority
requirements, provided that each of those losses arises from 'an inherent
defect in the Structures', and that the defect existed prior to practical
completion but was undiscovered at that time. It is important to
understand that 'inherent defects' are not any defect but have a restrictive
meaning under the policy wording. The 'Structures' for the purpose of the
policy are: external walls and claddings (including glazed curtain wallings
and non-load bearing facings and their fixings), floors and stairs, roofs
and roof structures, foundations, columns, beams and all other external

and internal load bearing elements essential to the stability of the property. Excluded from that definition are roof coverings and roofing materials, waterproofing membranes, windows and window frames and other non-load bearing elements of the building including floor coverings.

10.20   There are a number of exclusions to the cover including the failure to carry out any works recommended in the certificate of practical completion; any defect or other unsatisfactory feature which is the subject of a qualification by the independent consulting engineer appointed by the insurer to monitor the project throughout the design and construction stages; the use of the building for a purpose other than that for which it was originally designed or carrying any load in excess of the original safe design load; any defect which is discovered during the defects liability period, the remedying of which is a responsibility of the contractor under the building contract; subsidence, heave or landslip unless caused by an inherent defect in the foundations; faulty, inadequate or deficient weatherproofing or waterproofing (but this does not exclude destruction or damage to any parts of the building above ground level caused by damage which is covered); and a change in colour, texture, opacity or any ageing process or staining. There are other exclusions in addition to these.

10.21   It is an essential feature of this policy that an independent consulting engineer appointed by the insurer monitors the project through the design and construction stages; all that is at the expense of the insured and is in addition to the premium.

10.22   The excess under this policy will usually be a minimum of at least £25,000 increasing at years four and seven of the ten year policy, to reflect inflation. On the other hand, there is provision in the policy for inflation on the cost of rebuilding (in respect of which there will be an additional premium if inflation is running at a substantially higher level than that anticipated at inception of the policy). Under the standard policy, cover is provided in respect of loss of rent (being the actual amount of the reduction in the rent receivable by the insured arising out of an insured risk) and the additional expenditure necessarily and reasonably incurred in consequence of an insured risk solely to avoid or minimise the loss of rent (but not exceeding the amount of the reduction avoided by that expenditure).

10.23   Extensions are available to this policy to cover the cost of removal, temporary storage and re-installation of any of the insured property and other fixtures or fittings belonging to the insured (so as to permit the building to be repaired), and to cover, subject to the insurers' consent, space heating, ventilation and air conditioning systems, lifts and escalators, electrical distribution systems and (after a period of 12 months after the date of practical completion and initially for a five year period) weatherproofing and waterproofing.

Rights of subrogation against third parties are not waived and, indeed, **10.24** the insurer reserves the right to pursue recovery themselves, although they do say in their 'Technical Guide' that they will only exercise that right where they consider that the circumstances and the amount of the claim justify taking such action.

Some insurers will now consider cover on a ten year basis on completed **10.25** buildings, even where there has not been a technical appraisal on behalf of the insurer during the design and construction process; such insurance, where it is available, will usually be at a premium increased over and above the level which it would have been had there been appropriate earlier inspection.

POSSIBLE SOLUTIONS — THE FUTURE

Standard Forms of Collateral Warranty

There can be no doubt that the agreement of widely accepted standard **10.26** forms of collateral warranty would be of enormous benefit to tenants, property developers, funds, architects, engineers, contractors and sub-contractors. However, the wide diversity of forms that are needed to cater not only for the different parties, but also for the different methods of procurement makes this a very large task indeed. It may be beneficial for there to be standard clauses agreed which can then be incorporated into the different forms of agreement in relation to the different methods of procurement. On the other hand, the competing vested interests of the parties in trying to agree such terms make the task exceedingly difficult. It is to be hoped, however, that there will be some success in this direction as quickly as it can possibly be achieved.

The efforts of the BPF (in conjunction with the RIBA, ACE and RICS) **10.27** are much to be welcomed in this field.

Developments in Insurance

There can be no doubt that the insurance market is responding to the **10.28** difficulties created by the changing law of tort in relation to latent building defects and owners, purchasers and tenants. Premium levels on ten year non-cancellable building insurance appear to be becoming more competitive and the optional cover available is becoming wider; the fact that it is now sometimes possible, subject to insurers' approval in each case, to obtain insurance of this type on completed buildings is a helpful development.

10.29   Waiver of subrogation rights against third parties in ten year policies (as envisaged in BUILD) is now sometimes available on payment of an increased premium. That development is particularly to be welcomed. If it became more widespread, then the prospect of lengthy and costly litigation in relation to building defects would be substantially reduced.

10.30   A particular problem arising in relation to the Wren Insurance Association professional indemnity insurance scheme for architects is now likely to be met by new proposals from Wren (which have not yet been announced in detail but which are under consideration at the time of publication of this book). Wren will presently only permit their architect insureds to enter into collateral warranties where the damages recoverable are limited to the cost of the remedial works. In other words, the liability for the indirect consequential loss (loss of rent, removal of the tenant, the rent of other premises, returning the tenant to the original building and all the consequential costs including loss of profit) is not covered by the warranty. Wren are presently thought to be investigating the provision, in conjunction with collateral warranties acceptable to them, of property developers' insurance (ten year non-cancellable on the building), together with a consequential loss including loss of profits insurance for occupiers of the building. It is only when details of this scheme come to be announced that building owners and tenants will be able to see whether it provides more extensive and/or more competitive cover than that available from the other insurers in this field of business.

European Community Proposals

10.31   In February of 1990, Claude Mathurin, Ingénieur Général des Ponts et Chaussées, delivered his long-awaited final report to the Commission of European Communities in Brussels: 'Study of responsibilities, guarantees and insurance in the construction industry with a view to harmonisation at Community level'. The report follows a resolution adopted by the European Parliament on 12 October 1988 calling for the standardisation of contracts and controls in the construction industry and the harmonisation of responsibilities and standards governing after-sales guarantees on housing. The thinking behind this resolution was in part that if a developer in country A wished to retain an architect in country B and an engineer in country C with a view to a development in country D, to be constructed by a contractor from country E, then the internal market in the EEC should permit that to happen, by a rationalisation and a harmonisation of procedures. It is easy to appreciate, therefore, the monumental task that faced Claude Mathurin. Even the start of his consideration of these issues would have been daunting to most people: what is the relevant law in each of the countries of the Common Market in relation to these issues?

The final report is an extensive and complicated document, but insofar **10.32** as it can be summarised, the proposals are:

- to define the main functions of those involved in any construction project, especially the role of the principal designer
- harmonisation of building control
- standardisation of the responsibility of the various parties involved, from acceptance of the works and for a realistic and reasonable length of time, taking into account the durability of the works and the nature of the works
- a minimum five year guarantee of satisfactory completion and durability from builders
- effective protection for buyers of new or renovated houses against construction defects and damage, by means of high-quality insurance schemes (the existing English NHBC scheme receiving praise in this report)
- improvement of the relationships between the parties involved in construction.

The report itself acknowledges the difficulty of achieving such aims. **10.33** Although no final decision appears to have been made at the date of publication of this book, it is looking increasingly likely that the proposals of Claude Mathurin are going to be put on hold by the Commission save in respect of his recommendations in relation to housing. If that is the position, then a great many people in the construction industry and professions in Europe will be both surprised and disappointed (for further information see *Construction News*, 31 May 1990 and *New Builder*, page 18, 24 May 1990). It does not seem, therefore, that the hoped-for proposals that might do away with the need for collateral warranties will be forthcoming in England and Wales by reason of EEC requirements, at least in the foreseeable future.

Statutory Provisions

Two aspects where a statutory intervention might be considered are the **10.34** issues raised in the Department of Trade and Industry report 'Professional Liability' and the Defective Premises Act 1972. These are considered separately.

'Professional Liability'

10.35 In November of 1989, the Department of Trade and Industry published 'Professional Liability: Report of the Study Teams', a document which had been submitted to the DTI and the Department of the Environment as early as April of 1989. That report was under the chairmanship of Professor Andrew Likierman and is available from Her Majesty's Stationery Office. The Steering Group Report brings together the threads that emerged from separate reports prepared by the Auditors Study Team, the Construction Professionals Study Team and the Surveyors Study Team. The chairman of the Construction Professionals Study Team was Professor Donald Bishop and the members included architects, engineers, contractors, insurance underwriters, private and public sector clients, a quantity surveyor and a distinguished lawyer, Mr Donald Keating QC. The report of the Construction Professionals Study Team is a comprehensive analysis of the many liability issues that arise in construction. They concluded that doing nothing by way of reform would perpetuate what they perceived as a current erosion of trust and confidence. That in turn would lead to defensive tactics and 'risk shedding' becoming accepted practice, both impeding change and reducing efficiency and competitiveness. The Study Team went on to make a great many useful recommendations. Two of those recommendations were addressed to the government.

10.36 The first is that serious consideration should be given to altering the law on limitation of action to provide:

- a limitation period of ten years from the date of practical completion or effective occupation (staged if necessary) for negligence actions in tort and in contract (whether or not under seal)
- the ten year limitation period acting as a longstop extinguishing the right (c.f. Latent Damage Act 1986)
- redefinition of the term 'deliberate concealment' so that ordinary construction processes would not lead to an exception to the longstop
- consideration to be given to shortening the period allowed for launching contribution proceedings.

10.37 Secondly, the Study Team recommended that the government should give serious consideration to the alteration of the law on joint and several liability. Their recommendation was that the law in relation to joint liability should be altered for commercial transactions not involving personal injury and where the plaintiff's claim exceeds, say, £50,000. In such cases, it is recommended that a defendant whose actions are partly the cause of damage to the plaintiff, but such actions were not carried out jointly with another defendant, should be responsible to the plaintiff only for damage equivalent to that part of the plaintiff's loss fairly attributable to his or her breach.

It remains to be seen whether any government action whatsoever will be taken in relation to these recommendations. **10.38**

Amendments to the Defective Premises Act 1972
The name of this Act is a little misleading; it only applies to the provision of dwellings. The Act is considered at 4.24 above. **10.39**

The Act contains its own limitation period in respect of the duty, namely six years from the time when the dwelling was completed, but if further work is done to rectify the work that has already been done, then it is six years from the date when that further work was finished. **10.40**

If the government had the necessary will, the Law Commission could be asked to produce either a new draft statute along the lines of the Defective Premises Act or draft amendments to the Defective Premises Act to extend the scope of the Act to construction work in general. Such drafting would have to take into account at least the following points: **10.41**

(1) Who is to owe the statutory duty? This should be wide enough to encompass architects, engineers, contractors and sub-contractors.
(2) In respect of what type of buildings is the duty to be owed? Policy considerations might restrict the construction work definition. For example, what would be the position on civil engineering projects and government owned buildings?
(3) To whom is the duty to be owed? As in the present Act, there is no reason why the duty should not be owed to the person who ordered the construction work *and* every person who acquires an interest (legal or equitable) in the property.
(4) What should the nature of the duty be? It does seem that fit for habitation would probably not be the best type of duty to be owed in relation to building works generally for the simple reason that not all building works are intended to be fit for habitation; although not all building works are intended for occupation, it may be that that could be a suitable substitution for habitation in cases

where there is to be occupation. In any event, the duty should extend to carrying out the work in a professional manner with materials of good quality and in a good and workmanlike manner.

(5) There may well have to be limitations on the duty, such as the 'instructions' exception in section 1(2) of the present Act.

(6) What should the limitation period be? It does seem that the most likely contender would be ten years from practical completion, or from the date of completion of work to remedy defective work, whichever is the later. This would be consistent with the recommendations of the DTI 'Professional Liability Report' (see 10.36) and with the ten year non-cancellable building insurance policies (see 10.14 to 10.25).

10.42 It is an anomaly of our present law that certain types of duty are created by statute in respect of the provision of dwellings but not in respect of other types of construction; that anomaly is even more profound since the restriction of tortious duties in *D & F Estates Limited and Others* v. *The Church Commissioners for England and Others* and in *Murphy* v. *Brentwood District Council*. It cannot be in the public interest for there to be a continuation of the present explosion in the use of collateral warranties which can only, in the long term, create difficulties of legal interpretation, uncertainty and extensive legal costs for those involved in the consequential litigation on the warranties that have been and will be signed.

Appendix 1

BPF Form of Agreement for Collateral Warranty for funding institutions CoWa/F with General advice and Commentary on clauses

Reproduced with the kind permission of the BPF, ACE, RIBA and RICS, the joint copyright holders

Form of Agreement for

Collateral Warranty for funding institutions

CoWa/F

The forms in this pad are for use where a warra*
is to be given to a company providing finance fo*
proposed development. They must not in a*
circumstances be provided in favour of prospe*
tive purchasers or tenants.

General advice

1. The term 'collateral contract' or 'collateral warranty' is often used without due regard to the strict legal meaning of the phrase. It is used here for agreements with a funding institution putting up money for construction and development.

2. The purpose of the Agreement is to bind the party giving the warranty in contract where no contract would otherwise exist. This can have implications in terms of professional liability and could cause exposure to claims which might otherwise not have existed under common law.

3. The information and guidance contained in this note is designed to assist consultants confronted with a demand that collateral agreements be entered into.

4. The use of the word 'collateral' is not accidental. It is intended to refer to an agreement that is an adjunct to another or principal agreement, namely the conditions of appointment of the consultant. It is imperative therefore that before collateral warranties are executed the consultant's terms and conditions of appointment have been agreed between the client and the consultant and set down in writing.

5. The terms and conditions of the consultant's appointment may be 'under hand' or in the nature of a Deed and executed 'under seal'. In the latter case the length of time that claims may be brought under the agreement is extended from six years to twelve years.

6. This Form of Agreement for Collateral Warranty is designed for use under hand or under seal. It should not be signed under seal when it is collateral to an appointment which is under hand.

7. The acceptance of a claim under the consultant's professional indemnity policy, brought under the terms of a collateral warranty, will depend upon the terms and conditions of the policy in force at the time when a claim is made.

8. Consultants with a current policy taken out under the RIBA, RICSIS or ACE schemes will not have a claim refused simply on the basis that it is brought under the terms of a collateral warranty provided that warranty is in this form. In other respects the claim will be treated in accordance with policy terms and conditions in the normal way. **Consultants insured under different policies** must seek the advice of their brokers or insurers.

9. **Amendment to the clauses should be resisted.** Insurers' approval as mentioned above is in respect of the unamended clauses only.

Published by
The British Property Federation Limited
35 Catherine Place
London SW1E 6DY
Telephone: 071-828 0111

© The British Property Federation, The Association of Consulting Engineers, The Royal Institute of British Architects and The Royal Institution of Chartered Surveyors. 1990.

ISBN 0 900101 08 6

Commentary on clauses

Recitals A, B and C are self-explanatory and ne*
completion. The Consultant is described in the form as "T*
Firm". The following notes are to assist in understanding *
use of the document:

Clause 1
This confirms the duty of care that will be owed and furth*
that any obligation to the third party will be no greater th*
the obligations owed to the client. The terms "skill and ca*
or "skill, care and diligence" should accord with *
conditions of engagement.

Clause 2
As consultant it is not possible to give assurances beyo*
those to the effect that materials as listed will not *
specified. Concealed use of such materials by a contrac*
could possibly occur, hence the very careful restriction*
terms of this particular warranty. Further materials may *
added.

Clause 4
This obliges the consultant to ensure that all fees due a*
owing at the time the warranty is entered into have be*
paid.

Clause 5
This entitles the funding organisation to take over t*
consultant's appointment from the client on terms that *
fees outstanding will be discharged by the funding author*
(see Clause 7).

Clause 6
This affects the consultant's right to determine t*
appointment with the client in the sense that the fundi*
authority will be given the opportunity of taking over t*
appointment, again subject to the payment of all fees whi*
is the purpose of **Clause 7**.

Clause 8
This clause gives to the third party all the rights that th*
should reasonably expect in terms of use of drawings, etc*
does not give a licence to reproduce the designs contain*
within them other than for the purposes of the Developme*

Clause 9
This is a provision confirming that professional indemn*
insurance will be maintained in so far as it is reasonab*
possible to do so. Professional indemnity insurance is *
the basis of annual contracts and the terms and conditio*
of a policy may change from renewal to renewal.

Clause 11
This clause indicates the right of assignment by the fundi*
institution

Clause 12
This should be completed by identifying and agreeing t*
other parties required to sign similar agreements.

> N.B. The above advice and commentary is not intended to affect the interpretation of this Collateral Warranty. It is based on the terms of insurance current at the date of publication. All parties to the Agreement should ensure that terms of insurance have not changed.

Warranty Agreement   CoWa/F

Note

This form is to be used where the warranty is to be given to a company providing finance for the proposed development. Where that company is acting as an agent for a syndicate of banks, a recital should be added to refer to this as appropriate.

THIS AGREEMENT

is made the_____day of_____199_____

BETWEEN: –

(insert name of the Consultant)

(1) _____

of/whose registered office is situated at_____

_____("the Firm");

(insert name of the Firm's Client)

(2) _____

whose registered office is situated at_____

_____("the Client"); and

(insert name of the financier)

(3) _____

whose registered office is situated at_____

_____

("the Company" which term shall include all permitted assignees under this agreement)

WHEREAS: –

(insert description of the works)

(insert address of the development)

A. The Company has entered into an agreement ("the Finance Agreement") with the Client for the provision of certain finance in connection with the carrying out of

_____

_____

at_____

_____("the Development").

(insert date of appointment)

(delete as appropriate)

B. By a contract ("the Appointment") dated_____

the Client has appointed the Firm as [architects/consulting structural engineers/ consulting building services engineers/   surveyors] in connection with the Development.

(insert name of building contractor or "a building contractor to be selected by the Client")

C. The Client has entered or may enter into a building contract ("the Building Contract") with

_____

_____

for the construction of the Development.

NOW IN CONSIDERATION OF THE PAYMENT OF ONE POUND (£1) BY THE COMPANY TO THE FIRM (RECEIPT OF WHICH THE FIRM ACKNOWLEDGES) IT IS HEREBY AGREED as follows: –

(to reflect terms of the Appointment)

1. The Firm warrants that it has exercised and will continue to exercise reasonable skill [and care] [care and diligence] in the performance of its duties to the Client under the Appointment, provided that the Firm shall have no greater liability to the Company by virtue of this Agreement than it would have had if the Company had been named as a joint client under the Appointment.

(Delete where the Firm is the quantity surveyor)

[2. Without prejudice to the generality of Clause 1, the Firm further warrants that it has exercised and will continue to exercise reasonable skill and care to see that, unless authorised by the Client in writing or, where such authorisation is given orally, confirmed by the Firm to the Client in writing, none of the following has been or will be specified by the Firm for use in the construction of those parts of the Development to which the Appointment relates: –

    (a) high alumina cement in structural elements;

    (b) wood wool slabs in permanent formwork to concrete;

    (c) calcium chloride in admixtures for use in reinforced concrete;

    (d) asbestos products;

    (e) naturally occurring aggregates for use in reinforced concrete which do not comply with British Standard 882: 1983 and/or naturally occurring aggregates for use in concrete which do not comply with British Standard 8110: 1985.

(Further specific materials may be added by agreement)

    (f)

                                                                       ]

3. The Company has no authority to issue any direction or instruction to the Firm in relation to performance of the Firm's duties under the Appointment unless and until the Company has given notice under Clauses 5 or 6.

4. The Firm acknowledges that the Client has paid all fees and expenses due and owing to the Firm under the Appointment up to the date of this Agreement. The Company has no liability to the Firm in respect of fees and expenses under the Appointment unless and until the Company has given notice under Clauses 5 or 6.

5. The Firm agrees that, in the event of the termination of the Finance Agreement by the Company, the Firm will, if so required by notice in writing given by the Company and subject to Clause 7, accept the instructions of the Company or its appointee to the exclusion of the Client in respect of the Development upon the terms and conditions of the Appointment. The Client acknowledges that the Firm shall be entitled to rely on a notice given to the Firm by the Company under this Clause 5 as conclusive evidence for the purposes of this Agreement of the termination of the Finance Agreement by the Company.

6. The Firm further agrees that it will not without first giving the Company not less than twenty one days' notice in writing exercise any right it may have to terminate the Appointment or to treat the same as having been repudiated by the Client or to discontinue the performance of any duties to be performed by the Firm pursuant thereto. The Firm's right to terminate the Appointment with the Client or treat the same as having been repudiated or discontinue performance shall cease if, within such period of notice and subject to Clause 7, the Company shall give notice in writing to the Firm requiring the Firm to accept the instructions of the Company or its appointee to the exclusion of the Client in respect of the Development upon the terms and conditions of the Appointment.

7. It shall be a condition of any notice given by the Company under Clauses 5 or 6 that the Company or its appointee accepts liability for payment of the fees payable to the Firm under the Appointment and for performance of the Client's obligations under the Appointment including payment of any fees outstanding at the date of such notice. Upon the issue of any notice by the Company under Clauses 5 or 6, the Appointment shall continue in full force and effect as if no right of termination on the part of the Firm had arisen and the Firm shall be liable to the Company or its appointee under the Appointment in lieu of its liability to the Client. If any notice given by the Company under Clauses 5 or 6 requires the Firm to accept the instructions of the Company's appointee, the Company shall be liable to the Firm as guarantor for the payment of all sums from time to time due to the Firm from the Company's appointee.

8. The copyright in all drawings, reports, specifications, bills of quantities, calculations and other similar documents provided by the Firm in connection with the Development shall remain vested in the Firm but the Company and its appointee shall have a licence to copy and use such drawings and other documents and to reproduce the designs contained in them for any purpose related to the Development including, but without limitation, the construction, completion, maintenance, letting, promotion, advertisement, reinstatement and repair of the Development. The Company and its appointee shall have a licence to copy and use such drawings and other documents for the extension of the Development but such use shall not include a licence to reproduce the designs contained in them for any extension of the Development. The Firm shall not be liable for any such use by the Company or its appointee of any drawings and other documents for any purpose other than that for which the same were prepared and provided by the Firm.

(insert amount) (insert period)
9. The Firm shall maintain professional indemnity insurance in an amount of not less than pounds (£) for any one occurrence or series of occurrences arising out of any one event for a period of years from the date of practical completion of the Development for the purposes of the Building Contract, provided always that such insurance is available at commercially reasonable rates. The Firm shall immediately inform the Company if such insurance ceases to be available at commercially reasonable rates in order that the Firm and the Company can discuss means of best protecting the respective positions of the Company and the Firm in respect of the Development in the absence of such insurance. As and when it is reasonably requested to do so by the Company or its appointee under Clauses 5 or 6, the Firm shall produce for inspection documentary evidence that its professional indemnity insurance is being maintained.

10. The Client has agreed to be a party to this Agreement for the purposes of Clause 12 and for acknowledging that the Firm shall not be in breach of the Appointment by complying with the obligations imposed on it by Clauses 5 and 6.

11. This Agreement may be assigned by the Company by way of absolute legal assignment to another company providing finance or re-finance in connection with the Development without the consent of the Client or the Firm being required and such assignment shall be effective upon written notice thereof being given to the Client and to the Firm.

(insert names and/ or descriptions of other parties required to sign warranty agreements)
12. The Client undertakes to the Firm that warranty agreements in the Model Form CoWa/F published by the British Property Federation or in substantially similar form have been or will be entered into between_____

_____

_____

on the one hand and the Company on the other hand.

13. Any notice to be given by the Firm hereunder shall be deemed to be duly given if it is delivered by hand at or sent by registered post or recorded delivery to the Company at its registered office and any notice to be given by the Company hereunder shall be deemed to be duly given if it is addressed to "The Senior Partner"/"the Managing Director" and delivered by hand at or sent by registered post or recorded delivery to the above-mentioned address of the Firm or to the principal business address of the Firm for the time being and, in the case of any such notices, the same shall if sent by registered post or recorded delivery be deemed to have been received forty eight hours after being posted.

(Alternatives: delete as appropriate)

AS WITNESS the hands of the parties the day and year first before written.

(These must only apply if the Appointment is under seal)

IN WITNESS whereof the partners in the Firm have hereunto set their hands and seals and the Common Seals of the Company and the Client were hereunto affixed the day and year first before written.

IN WITNESS whereof the common seals of the parties were hereunto affixed the day and year first before written.

(NB: All parties referred to in Clause 12 must use the same attestation clause)

Appendix 2

The Royal Incorporation of Architects in Scotland Duty of Care Agreement

Reproduced with the kind permission of the RIAS who hold the copyright.

Note: This Agreement is drafted under the law of Scotland and is not to be used for warranties under English law.

THE ROYAL INCORPORATION
OF ARCHITECTS IN SCOTLAND

AGREEMENT

between

_____

(hereinafter referred to as "the Client")

and

_____

(hereinafter referred to as "the Company")

and

_____

(hereinafter referred to as "the Architect")

Approved for use in Scotland by Scheme Underwriters
to RIAS Insurance Services Limited

CONSIDERING THAT the Client
and the Company have entered into
an Agreement ....

....

....

....

regarding ....

....

....

....

....

at..

....

(hereinafter referred to as "the
Development"):

FURTHER CONSIDERING that
the Client has appointed the
Architect as ....

....

....

in connection with the Development
and has entered into or is about to
enter into a Building Contract with

....

for the erection of the Development:

FURTHER CONSIDERING that in
addition to the fees and expenses
referred to in Clause Four thereof the
Client has paid the Architect the sum
of ....

....

pounds (£...)
receipt of which the Architect hereby
acknowledges.

THEREFORE the Client, the
Company and the Architect HAVE
AGREED and DO HEREBY
AGREE as follows:

ONE The Architect undertakes that he has
exercised and will continue to
exercise all reasonable skill and care
in the performance of his duties to the
Client in connection with the
Development.

TWO The Architect undertakes that he has
not now and will not in the future
specify for use in the Development
any of the following materials:-

a High alumina cement in structural
elements

b Wood wool slabs in permanent
formwork to concrete or in
structural elements.

c Calcium chloride in admixtures for
use in reinforced concrete.

d Asbestos products.

e Aggregates for use in reinforced
concrete which do not comply with
British Standard Specification
882:1983 and aggregates for use in
concrete which do not comply with
the provisions of British Standard
Specification 8110:1985.

THREE The Company shall not be entitled to
issue instructions to the Architect in
relation to the Development unless
and until the Company has given
notice to the Architect under Clause
FIVE.

FOUR The Architect acknowledges that he
has received from the Client all fees
and expenses inclusive of VAT due to
him up to the date of this Agreement
and that the Company shall have no
liability for fees due to the Architect
until the Architect has received the
notice referred to in Clause FIVE
hereof.

FIVE In the event of termination of the
Agreement between the Client and
the Company, the Architect shall, on
receipt of written notice from the
Company, accept instructions solely
from the Company in respect of the
carrying out and completion of the
Development: provided that any such
notice shall state that the Company
accepts liability for payment of all fees
due to the Architect and for
performance of the Client's
obligations under the Architect's
Appointment. Receipt of such notice
by the Architect, provided a copy is
sent by the Company to the Client,
shall be accepted by all parties to this

Agreement as conclusive evidence of the termination of the agreement between the Client and the Company.

SIX The Architect shall not, without first giving the Company not less than Fourteen days' written notice, terminate his appointment with the Client or treat his appointment as having been repudiated by the Client or discontinue the performance of any duties under the appointment: provided that the Architect's right to terminate, repudiate, or discontinue performance as the case may be, shall cease on receipt of the notice from the Company referred to in Clause FIVE.

SEVEN The Architect shall retain the copyright in all Drawings, Reports, Specifications, Bills of Quantities, Calculations and other similar documents provided by him in connection with the Development but provided the Architect's whole fees and disbursements to date have been paid and subject to the conditions of Clause 3.16 of the Architect's Appointment the Client and the Company shall have a licence to copy and use such Drawings and other documents and to reproduce the designs contained in them for any purpose relating to the Development including the construction, completion, maintenance, letting, promotion, advertisement, reinstatement, repair or extension of the Development, but for no other purpose.

EIGHT The Architect shall use his best endeavours to maintain professional indemnity insurance for the Development of not less than................

Insert agreed sum

....

pounds (£..)
for any one occurrence or series of occurrences arising out of any one

If left blank, the period will be not more than five years

event for a period of .... years from the Date of Practical Completion of the Development as defined in the Building Contract: provided always that the annual premium on such insurance is not unreasonable in the context of the Architect's annual turnover from year to year. The Architect shall immediately inform the Company when he considers such insurance cannot reasonably be obtained.

The Architect shall as and when required by the Company produce documentary evidence of the existence of such professional indemnity insurance.

NINE The Company accepts that the Architect shall, as a result of this Agreement, have the same but no greater liability to the Company than he does to the Client.

TEN Either the Client or the Company shall be entitled to assign or transfer their rights under this Agreement to any party acquiring an interest in the Development or any part thereof, from them or any one of them, within a period of 5 years from the date of issue by the Architect of the Certificate of Practical Completion: provided that in no circumstances shall the Client or the Company be entitled to assign or transfer their interests in this Agreement after the expiry of the said period of 5 years. Nothing in this clause shall permit any party acquiring such right as assignee or transferee to enter into any further assignation or transfer to anyone acquiring an interest in the Development or any part thereof from him.

Any assignation entered into by the Client or the Company under the terms of this Clause shall forthwith be intimated in writing to the Architect.

ELEVEN This Agreement and any assignation or transfer following thereon in accordance with Clause TEN hereof shall automatically be null and void and neither the Client, the Company nor the Architect shall be bound by any provisions therein if the Development is used for a purpose for which it was not designed, unless the Architect has confirmed in writing that such purpose will not affect the structure or stability of the Development.

TWELVE   Any notice which requires to be sent in accordance with the Agreement shall be sent by recorded delivery to the addressee's address as stated herein and shall be deemed to have been received Twenty-four hours after being posted IN WITNESS WHEREOF these presents are executed as follows:-

The COMMON SEAL of the Client was

hereunto affixed at ....

on the..

day of ....

Nineteen hundred and...

in the presence of: Affix Seal here

Director...

Secretary ....

(Or otherwise as stipulated in Memorandum

and Articles of Association)

The COMMON SEAL of the Company was

hereunto affixed at...

on the..

day of ....

Nineteen hundred and...

in the presence of:-   Affix Seal here

Director...

Secretary ....

(Or otherwise as stipulated in Memorandum

and Articles of Association)

SIGNED by the above-named Architect

at ....

on the ....

day of ....

Nineteen hundred and...

before these witnesses:-

Signature .... (Witness)

Address ....

....   (Signature of architectural practice)

Occupation ....

Signature .... (Witness)

Address ....

....

Occupation ....

(N.B. This testing clause has been drafted on the assumption that the Client and the Company are Limited Liability Companies and the Architect a partnership: if this is not the case, suitable amendments will have to be made.)

Table of Cases

Note – the following abbreviations of Reports are used:

AC — Law Reports, Appeal Cases
All ER — All England Law Reports
ALR — Australian Law Reports
Asp MLC — Aspinall's Maritime Law Cases
Atk — Atkins
Beav — Beavan
Bing — Bingham
BLR — Building Law Reports
C & P — Carrington & Payne Reports
Ch. — Law Reports, Chancery Division
CILL — Construction Industry Law Letter
CL — Current Law
CLR — Commonwealth Law Reports
CLY — Current Law Year Book
Com Cas — Commercial Cases
Com.LR — Commercial Law Reports
ConLR — Construction Law Reports
Const.LJ — Construction Law Journal
Cro. Eliz — Croke, Time of Elizabeth I
Deac & Ch — Deacon & Chitty
DeGF and J — De Gex Fisher & Jones
DLR — Dominion Law Reports
DPC — Davies' Patent Cases
Dyer — Dyer
EG — Estates Gazette
Ex. — Law Reports, Exchequer Division
FTLR — Financial Times Law Reports
Gal & Dav — Gale & Davison
ICLR — International Construction Law Review
IR — Irish Reports

JP — Justice of the Peace
Jur [NS] — Jurist Reports [New Series]
KB — Law Reports, King's Bench Division
KIR — Knights Industrial Reports
LGR — Local Government Reports
LJ — Law Journals
LJCP — Law Journal, Common Pleas
LS Gaz — Law Society Gazette
LT — Law Times Report
LLR — Lloyd's List Law Reports
Lloyd's Rep. — Lloyds Law Reports
Mont & A — Montague & Ayrton
MOO & P — Moore & Payne
My. Cr. — Myine & Craig
New LJ — New Law Journal
NSW — New South Wales Law Reports
PC — Privy Council
PD — Law Reports, Probate Division
QB — Law Reports, Queen's Bench Division
R & IT — Rating & Income Tax
RR — Revised Reports
Russ — Russell
SC — Session Cases
SLT — Scots Law Times
Sol. J — Solicitors' Journal
SMLC — Smith's Leading Cases
WLR — Weekly Law Reports
WR — Weekly Reporter
Y and CEX — Younge & Collyer (Exchequer, Equity)

References are to section numbers

Table of Statutes & Statutory Instruments

References are to section numbers

Index

References are to section numbers

claims under a contract 5.2
co-defendants, between 5.47, 5.48
copyright 8.45–8.48, 9.59–9.62
costs, legal 9.7, 9.8

damages *see also*, measure,
    mitigation *and* causation
    apportionment of 5.44–5.46
    ascertainment, difficulties of 5.27,
      5.28
    assessment, date of 5.26
    assignees, entitlement to 5.29–5.43
    betterment 5.24, 5.25
    consequential losses 5.3, 5.39–5.43,
      7.27
    direct costs 5.30–5.38
    economic loss 5.3
    future damages 5.2
    general damages 5.2
    liquidated and ascertained 5.3
    nature of 5.1
    special damages 5.2
delay
    provision of design, in 8.81–8.83
deleterious materials 8.34–8.44,
    9.42–9.44
design 8.19–8.29, 9.38–9.41
developer, interests of 9.15–9.21
disputes, resolution of 8.84–8.87
DTI Report on Professional Liability
    10.34–10.38
duty of care letters 8.17
dwellings 4.24–4.26

ESA/1 8.24
European Community, proposals of
    10.31–10.33
exclusion clauses 8.9–8.14, 8.72–8.77
execution 5.53, 9.12–9.13
    under hand 5.53, 8.89, 9.78
    under seal 8.89, 9.52, 9.55, 9.78
expectation interest
    meaning of 5.14

fitness for purpose *see also*
    reasonable skill and care
    4.14–4.23
    design and build contractors
      4.21–4.23
    professional indemnity insurance
      and 7.30–7.32
    warranties, in 8.30–8.31
fund
    collateral warranties, from whom
      6.28
    contents of collateral warranty, for
      6.29–6.30, 9.22–9.25
    problems of 6.26–6.27

indemnity provisions 8.80
inspection duties
    architects 8.29
    engineers 8.29
insurance, *see also* professional
    indemnity insurance 10.14–10.25
    BUILD 10.15–10.17
    building defects 10.18–10.25
    developments in 10.28–10.30
insurers
    changing of 7.36

JCT 80 9.18, 9.23
JCT IFC 84 9.23
JCT 'With Contractor's Design'
    8.27, 8.28, 9.20, 9.24
    architect's warranty 8.28, 8.29
    engineer's warranty 8.28, 8.29
JCT Works Contract/3 8.24

law reform, *see* reform of law *and*
    other solutions
lease
    landlord to bear repairing cost 6.3,
      10.2–10.9
    repairing covenant of tenant
      6.2–6.9
liability
    limiting 8.9–8.14, 8.72–8.77